THE SELLING OF BRITISH TELECOM

The Selling of British Telecom

KARIN NEWMAN

HOLT, RINEHART AND WINSTON
London · New York · Sydney · Toronto

Holt, Rinehart and Winston Ltd: 1 St Anne's Road,
Eastbourne, East Sussex, BN21 3UN

British Library Cataloguing in Publication Data

Newman, Karin
The Selling of British Telecom.

1. British Telecom 2. Telecommunication systems
—Great Britain 3. Privatization—Great Britain
I. Title
384'.065'41 HE8095

ISBN 0-03-910639-x

Typeset by Paston Press, Norwich.
Printed in Great Britain by Butler and Tanner Ltd, Frome.

Last digit is print no: 9 8 7 6 5 4 3 2 1

Preface

The sale of British Telecom was not the first measure of privatisation undertaken since 1979 but it is to date by far the largest and, by virtue of its daunting scale, it has also been the most demanding marketing exercise yet attempted in the UK. It is the sale of British Telecom to the retail sector which is the principal subject of this book. It was a collaborative effort, shared by Government departments, BT, banks, stockbrokers, accountants, lawyers and marketing specialists. In the process the techniques of planning and executing an unprecedented campaign were evolved. It deployed the whole modern armoury of marketing strategies and tactics. Large-scale and continuous market research, sophisticated audience analysis and versatile use of advertising media were coupled with the largest direct-mail campaign the world has yet seen. About 42.5 million items of print and film were distributed to the UK audience, direct mail achieved an outstanding conversion rate of 61 per cent and 2.3 million people (including BT employees) eventually took up their allotment.

An exercise of this magnitude has its own intrinsic interest, but it also has lessons for the future. Indeed, the success of the marketing of British Telecom shares seems to have broken through remaining inhibitions about the sale of public assets and may well have laid the foundations for the wider share ownership which its planners envisage. The Chancellor of the Exchequer announced on 12 November 1985 that British Gas, British Airways and Thames Water would be among the assets to be sold to the British public. 'The increased pace of privatisation means that the proceeds from this programme will rise substantially from £2.5 billion this year [1985] to £4.75 billion in each of the next three years.' Thus there will certainly be other and perhaps bigger flotations in the future but those who undertake them will find that the lessons of British Telecom are worth studying. There are features of this story which should prove instructive to promoters of new issues and interest professional practitioners as well as students of marketing.

The flotation of British Telecom was not a parochial matter. It had an important international dimension. Following the example of AT & T in the USA and the success of British Telecom here the indications are that privatisation of public enterprises is being contemplated in many industrialised and third world countries—France, Japan, Denmark, the Netherlands and Portugal as well as Brazil, Thailand and Bangladesh.

Many individuals and institutions have helped in the telling of this story and I am indebted to all those who have freely provided me with illustrations and material. It would not have been possible without the co-operation of British Telecom, the Department of Trade and Industry, Dewe Rogerson, MORI, Dorland Advertising and Burrup Mathieson. I am indebted to Roddy Dewe, who prompted me to undertake this book, and I must thank particularly John King, Peter Young and Peter Carver, all of British Telecom, for much practical help and advice. James Hudson of Dorland Advertising has been most generous with his time and Bob Worcester of MORI has given me invaluable encouragement. Anthony Carlisle's critical comments were salutary and I am indebted to others who do not wish for specific acknowledgement. Needless to say, none of those named or unnamed carry responsibility for any mistakes which remain.

Contents

Chapter 1

Political and Parliamentary Background

Although the policy of privatisation did not figure in the Conservative Party Manifesto of 1979, the Secretary of State for Industry (then Sir Keith Joseph) confirmed the Government's intention in a speech to the party conference of October:

> We have to manage the nationalised industries that we have inherited. We have to reduce the range of such industries by selling part of the shares of some to the public. We agree with the Bow Group in its publication today* that we should wherever possible introduce competition into the nationalised industries.

Sir Geoffrey Howe's first budget of June 1979 had already specified the objective of wider share ownership and 1980 saw the beginning of the process of privatisation with the British Aerospace Bill. In 1981 the Transport Act enabled British Rail to sell off its subsidiaries, and set up a holding company to control the reconstituted British Transport Docks Board. The same year produced the British Telecommunications Act which, as will be seen, was the necessary preliminary to privatisation.

The philosophy of this programme was amplified by Sir Keith Joseph in April 1981:

> We came to office convinced that the structure of the nationalised industries contributed to the national malaise . . . in all too many cases, particularly when the nationalised industry commanded a monopoly, those concerned did not see themselves as living under the healthy necessity of satisfying the customer in order to survive; they had no incentive to cut costs to beat competitors; they were free of the risk of liquidation . . . Such was our diagnosis: what was our aim? Our aim is to abate inflation and to create a prospering social market economy—that is, a mainly free enterprise economy.

* Fielding, J. (1979) *Public Monopoly and Private Enterprise*. Bow Group.

PRIVATISATION IN PRACTICE

Since the privatisation programme began in February 1981 the following principal enterprises have been transferred wholly or partly to private ownership:

1. British Aerospace, one of the largest aerospace organisations in the world, with turnover at the time of its sale of over £1000 million and pre-tax profits of about £50 million.
2. Cable and Wireless, a major international telecommunications organisation, which has been the subject of two sales of shares, with turnover at the time of the first sale of almost £300 million and pre-tax profits of about £60 million.
3. Amersham International, a supplier of radio chemicals, formerly part of the Atomic Energy Authority, with an annual turnover at the time of sale of about £60 million and pre-tax profits of about £8 million.
4. Britoil, the oil and gas production interests of the British National Oil Corporation, responsible for over 5 per cent of North Sea oil production at the time of its sale with pre-tax profits of about £470 million.
5. Associated British Ports, formerly the British Transport Docks Board, which owns and operates 19 ports in the United Kingdom. Like Cable and Wireless, ABP has been the subject of two sales of shares. At the time of the first sale, turnover and pre-tax profits were about £150 and £12 million respectively.
6. Enterprise Oil, the oil and gas interests of the British Gas Corporation, with forecast profits for the year of its sale of £113 million and net assets of £117 million.
7. Jaguar, a well known car manufacturer and a subsidiary of British Leyland, now BL, which is almost wholly owned by the Government. In 1983 it had a turnover of £472 million and pre-tax profits of £50 million.
8. The National Freight Corporation, which was bought by its management and employees with institutional financing, with net assets of £90 million and trading revenue in 1981 of £434 million.
9. British Telecom, the principal supplier of telecommunication services in the UK, valued at the time of sale at £7800 million.

For a fuller picture of the privatisation programme to date, see Table 1.1. The disposal of state industrial and commercial assets, raising over £7 billion, was not only sanctioned by Parliament but appears to have general support among the electorate. Successive attitudinal surveys by Market & Opinion Research International (MORI) have consistently revealed that the privatisation policy in general has been favoured by the population, and the decision to privatise British Telecom in particular was included in the Tory Party Manifesto of 1983 on which the Conservatives were returned with a record majority. From Figure

Table 1.1 *Privatisation in Britain, 1979–85*

Company	Business	Date of sale	Remaining Government holding %	Amount raised £m
British Petroleum	Oil	Oct 1979		276
		Jun 1981		8
		Sep 1983	31.7	543
British Aerospace	Aerospace	Feb 1981	49.4	150
		May 1985	Nil	363
British Sugar Corporation	Sugar refiner	Jul 1981	Nil	44
Cable and Wireless	Telecommunications	Oct 1981		182
		Dec 1983	23.1	263
Amersham International	Radiochemicals	Feb 1982	Nil	64
National Freight Company	Road haulage	Feb 1982	Nil	5
Britoil	Oil	Nov 1982	48.9	627
		Aug 1985	Nil	450
Associated British Ports	Seaports	Feb 1983		46
		Apr 1984	Nil	51
International Aeradio	Aviation communications	Mar 1983	Nil	60
British Rail Hotels	Hotels	Mar 1983	Nil	51
British Gas (Wytch Farm)	Oil	May 1984	Nil	82
Enterprise Oil	Oil	Jun 1984	Nil	380
Sealink	Harbours and ferries	Jul 1984	Nil	66
Jaguar	Cars	Jul 1984	Nil	297
British Telecom	Telecommunications	Nov 1984	49.8	3916

Source: *The Economist*, 23 February, 1985 (except for Britoil and British Aerospace 1985)

1.1 it can be seen that there was a marked swing of opinion during the 'winter of discontent' of 1974–75 and that thereafter Mrs Thatcher's call to roll back the frontiers of the state has been closely in tune with public opinion.

PHASE ONE—LIBERALISATION OF TELECOMMUNICATIONS

In September 1979, in the context of the rapid growth of telecommunications and of information technology in America, Sir Keith Joseph announced a Government review of British Telecom's monopoly in the telecommunications network, the supply of terminal equipment, and the provision of services to third parties using British Telecom's circuits. As a result of the review, Sir Keith in a statement to the House of Common on 21 July 1980 announced the Government's intention to liberalise the telecommunications market.

The consequent British Telecommunications Act received the Royal Assent in July 1981. This, the first of two measures which have transformed the telecommunications industry, is notable for inaugurating an autonomous British Telecom, and for introducing competition with British Telecom itself.

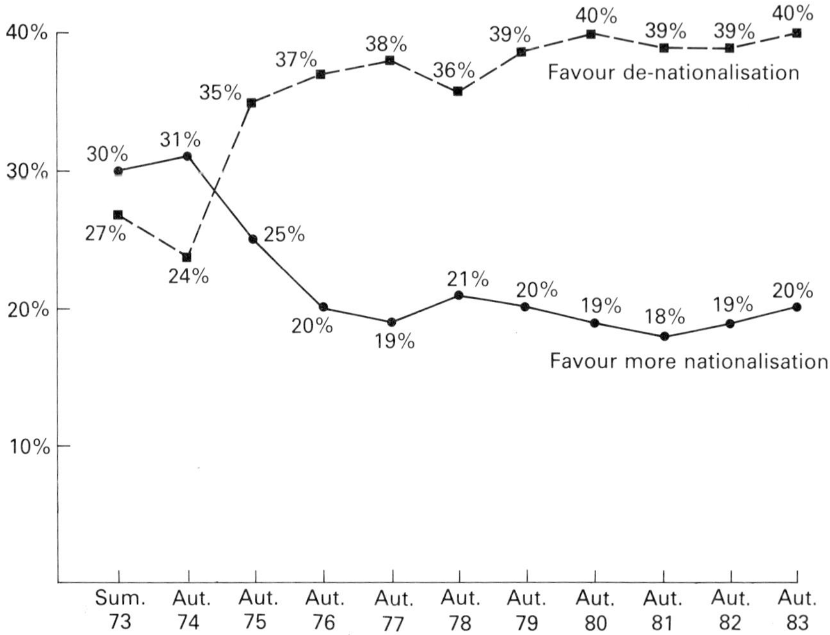

Figure 1.1 *Overall attitude to nationalisation—trends. Source: MORI.*

Clause 1 split telecommunications activity from the Post Office and set up a new corporation—the British Telecom Corporation. Clause 2 modified British Telecom's exclusive privilege to operate the domestic telephone network by giving the Government powers to license competitors. In October 1981, Mercury became the chosen child of telecommunication deregulation. It obtained a 25-year renewable licence to operate a national and international digital network to compete against British Telecom's trunk traffic; it will be the sole competitor until at least 1990. Mercury had been set up early in 1981 by British Petroleum, Barclays Merchant Bank and Cable and Wireless specifically to enter the business segment of long-distance communication with a customised service to corporate clients by establishing the planned programme of an optical fibre network in a figure of eight from London in the south east to Bristol in the south west, and then through Birmingham in the Midlands, to Leeds in north-central England and to Liverpool in the north west. London and Birmingham were to be linked by microwave. Mercury's licence also permitted it to interconnect with the British Telecom network and enter the profitable international communications sector to Europe and the United States. The Act, furthermore, brought to an end the monopoly British Telecom had previously enjoyed in the provision of telecommunications equipment. Only the provision of a user's first telephone remained its prerogative. The equipment market, ranging from telephones to private automatic branch exchanges

(PABXs), was thus opened up with only one proviso, that suppliers could only attach their equipment to BT's network provided it met with approval either from the Secretary of State or the British Approvals Board for Telecommunications. And finally, the 1981 Act enabled BT's telephone lines to be used by others to provide value-added services, for instance, data processing and storage, and answering facilities, where BT previously had a monopoly. The thrust of the statute was towards establishing long-term competition. The 1981 Act removed British Telecom's statutory monopoly in the network and equipment market and endowed the Secretary of State with powers to licence competitors. As such the British Telecommunications Act 1981 inaugurated a major potential change in British Telecom's environment.

Within the space of a year the changes unleashed by the 1981 Act became visible. Mercury's plans—if not their actual execution—to provide a service to business customers during 1983, were well advanced. Increasingly, the British Telecom network was being used by other firms to provide value-added services and the private sector was beginning to sell its telephone handsets directly to the public. At the same time British Telecom's new management was responding positively to incipient competition with an efficiency drive that dramatically reduced the waiting list for a telephone and reduced the time taken to repair faults. In addition a major investment drive had begun to introduce new digital equipment into the network, and to establish the System X exchanges. British Telecom also began an ambitious programme of new product offerings, including the use of microwave radio in the City of London overlay.

However, within a year of the British Telecommunications Act the telecommunications industry entered another stage in the development of Government policy. A new Secretary of State, Patrick Jenkin, in a statement to the House of Commons on Monday 19 July 1982, announced that the Government intended to privatise British Telecom. It is perhaps worth quoting his statement to the House, based on the White Paper, *The Future of Telecommunications in Britain*, in which the reasons for this dramatic step in policy are eloquently stated:

> It is the Government's aim to promote consumer choice. Wherever possible, we want industrial and commercial decisions to be determined by the market and not by the State. We believe that consumer choice and the disciplines of the market lead to more stable prices, improved efficiency and a higher quality of service.
> . . . As a nationalised industry British Telecom does not have access to financial markets. Its borrowing is controlled by Government and counts against the Public Sector Borrowing Requirement.
> . . . Monopoly power has allowed British Telecom to raise prices to finance investment without doing all that could be done to increase efficiency. Around 90 per cent of British Telecom's investment programme, about £2200 million this year, has been self-financed . . . by customers. As a result charges have risen steeply while investment is not enough.

For these reasons, the Government intended to introduce legislation after the next election which would convert the corporation into a public limited company under the Companies Acts, and offer 51 per cent of the company's shares to the open market. He assured his audience that:

> These proposals follow naturally from the liberalising measures passed by the House last year. It would make no sense to stop halfway . . . It cannot be right that British Telecom should remain subject to the web of Government interference and controls which are the inevitable lot of an industry which enjoys the privilege of Exchequer finance.

After the sale, multifaceted benefits were to be expected. For the Government there would be less administration since commercial decisions would be the preserve of the company and, more significantly, British Telecom would be removed from the public sector and Treasury control. British Telecom would no longer look to the Government for finance but to the private capital market. In effect, the transformation of British Telecom's status and share flotation would subject it to the disciplines of the market place, the judgement of its shareholders and to the demands of its customers.

However, the Government recognised that, despite the liberalisation of the market since 1981, the plan to float British Telecom as one entity rather than splitting it up into smaller independent units meant that British Telecom would continue to dominate the British market for many years. Therefore a system of regulation would be introduced to balance the various interests of suppliers, customers, employees and investors. A regulatory body, the Office of Telecommunications, modelled on the Office of Fair Trading, would be formed to ensure fair competition. Legislation would also end British Telecom's remaining exclusive privileges.

One of the most interesting questions in this story is, why did the Government mature this policy halfway through its term of office and announce the privatisation of British Telecom in 1982? The Conservative Manifesto of 1979 said nothing on the privatisation of British Telecom. The wish to roll back the frontiers of the state related primarily to the undoing of contentious nationalisations such as that of Cable and Wireless, nationalised in 1946, and of British Aerospace, nationalised more recently under the Aircraft and Shipbuilding Industry Act of 1977, whose return to the private sector was announced almost immediately after the Conservatives assumed office. Furthermore, the British Telecommunications Act 1981, which separated British Telecom from the Post Office and introduced limited competition, had made no reference to the privatisation of British Telecom. And yet, a year later, Patrick Jenkin announced a sharp change in policy direction—from limited competition and liberalisation to 51 per cent privatisation. What were the inducements to this unheralded development?

A number of factors may have contributed to the decision and timing of the announcement. Given the polarity of British politics at the time, there is the matter of ideological preferences. Whilst Labour stood for nationalisation, the Tories at the other end of the political spectrum expressed an extreme political dislike of state ownership, a dislike which in 1982 bordered on hatred of nationalised industries based on a perception of inefficiency and freedom from market forces. In tune with their free market philosophy there was also the certain belief that efficiency would be better gained from market forces than ineffective central state direction. Perhaps the feeling was best expressed when Norman Tebbit observed, shortly after his appointment as Secretary of State for Industry, 'The nationalised industries have been letting us down for years'.

The twin quest of efficiency and competition, however, related to a third element and that was the financial picture. Efficiency and competition, it was thought, would also reduce the drain on state finance and thus, alleviate the major problem of this monetarist Government. In 1982 the financial picture required some crucial decisions to be made in the way the nationalised industries were to be financed if public expenditure was to be contained. The cost of the nationalised industries threatened the public sector borrowing requirement limits. Almost unanimously chairmen of the nationalised industries were demanding more money to finance their major modernisation programmes and the sums required exceeded the strict cash limits imposed by the Treasury. In 1981, British Telecom alone persuaded the Government of the merits of its major investment programme and was allowed to seek external finance. Consideration was given to the introduction of a Government-backed 'Buzby Bond' which would have allowed British Telecom to raise possibly £100–£200 million in the City of London.

However, the financial question was never far away and was put into sharp relief in 1981 when statistics, newly published, showed that Government grants, loans and write-offs for the nationalised industries totalled almost £5 billion in 1980, equivalent to half the value of all the notes and coins in circulation. Perhaps even more crucial were the public-spending estimates published just after the Budget in March 1982 which forecast that the nationalised industries between them would need the huge sum of £5.3 billion of public support until the end of the Government's life. The amount itself may have been bad enough, but compared with an earlier estimate of only £800 million, it might have been a dramatic shock, providing the essential catalyst in the decision to privatise British Telecom, the most ambitious target to date. In Spring 1982 Patrick Jenkin killed off the 'Buzby Bond' idea; he announced the privatisation of British Telecom in July and introduced the first Telecommunications Bill four months later.

PHASE TWO—PRIVATISATION

In spite of Jenkin's statement, 'It is our intention, *after the next election* (my italics) to offer up to 51 per cent of the shares on the market in one or more tranches' a Bill to privatise British Telecom was printed on 17 November 1982. The latest date for a general election was May 1984. The purpose of the Bill, according to its long title is:

> To provide for the appointment of a Director General of Telecommunications and to confer functions on the Director General so appointed; to abolish the exclusive privilege with respect to telecommunications conferred on British Telecommunications by section 12 of the British Telecommunications Act 1981 and to make new provision with respect to the provision of Telecommunication services; to make provision, in substitution for the Telegraph Acts 1863 to 1916, for the matters there dealt with and related matters to provide for the vesting of all property, rights and liabilities of British Telecommunications in a company nominated by the Secretary of State and the subsequent dissolution of British Telecommunications; to make provision with respect to the finances of that company; to amend the Wireless Telegraphy Acts 1949 to 1967 and to make further provision for facilitating enforcement of those Acts; and for connected purposes.

Clause 1 empowered the Secretary of State to appoint the Director General, who in turn might appoint the staff of the Office of Telecommunications. It was expected that the staff of OFTEL would number about 50. Clause 2 abolished BT's exclusive privilege to run telecommunications systems. Clause 3 provided guidelines for the Secretary of State and the Director General, the first part of which may usefully be reproduced in full:

> In exercising the functions assigned to them by this Part the Secretary of State and the Director shall take into account all matters which appear to them in the particular circumstances to be relevant and, among other things, shall have regard to—
>
> (a) the need to secure so far as practicable the provision throughout the United Kingdom of such telecommunication services as satisfy all reasonable demands for them including, in particular, public call box services, emergency services and services in rural areas;
>
> (b) the need for persons supplying telecommunication services to achieve a return on capital employed sufficient to enable them to finance the supply of such services as are mentioned in paragraph (a) above and the development and use of new techniques;
>
> (c) the desirability of promoting the interests of consumers, purchasers and other users of telecommunication services and telecommunication apparatus in the United Kingdom in respect of the prices charged for them and in respect of their quality and variety of services and apparatus supplied;
>
> (d) the desirability of maintaining and promoting effective competition between persons supplying telecommunication services or telecommunication apparatus in the United Kingdom; and

(e) the desirability of maintaining and promoting competitive activity in markets outside the United Kingdom on the part of persons supplying telecommunication services, and persons producing and supplying telecommunications apparatus, in the United Kingdom.

Clauses 4 to 7 required all operators of telecommunication systems (with certain exceptions) to be licensed. A licence might be granted by the Secretary of State after consultation with the Director General, or by the Director General with the consent of the Secretary of State. Clause 8 made provisions for such licences to include certain conditions, including the payment of fees, and obligations to provide services. This Bill was subsequently amended by the House of Commons and had just begun its committee stage in the House of Lords, when it was lost by the dissolution of Parliament for the election of May 1983. After the election, the Bill was re-introduced and was essentially the same as the one given a third reading by the Commons and a second reading by the Lords in April 1983.

Clause 1 remained the same with the creation of OFTEL and the Director General. Clause 2 remained substantially the same and was concerned with the abolition of British Telecom's monopoly or exclusive privilege and paved the way for the establishment of the new licensing and regulatory regime. In fact, the exclusive privilege in practice no longer existed since the Secretary of State had already granted a licence to Mercury and introduced competition in the supply of telephone apparatus under powers from the 1981 Act. Clause 3 was the foundation of the regulatory machinery establishing broad policy guidelines within which the Secretary of State and Director General must operate when exercising their regulatory powers. Its importance was recognised during the Parliamentary stages of the first Bill and it was debated and refined by the Commons standing committee and during the Commons report stage. Consequently the clause has been improved and strengthened with the following additional points:

(a) to promote efficiency and economy on the part of such persons;
(b) to promote research into and the development and use of new techniques by such persons;
(c) to encourage major users of telecommunication services whose places of business are outside the United Kingdom to establish places of business in the United Kingdom;
(d) to promote the provision of international transit services by persons providing telecommunication services in the United Kingdom;
(e) to enable persons providing telecommunication services in the United Kingdom to compete effectively in the provision of such services outside the United Kingdom;
(f) to enable persons producing telecommunication apparatus in the United Kingdom to compete effectively in the supply of apparatus both in and outside the United Kingdom.

Consequently the broad sweep of the Clause now provided a balance between the telecommunications interests of the nation, providers, users of services, and suppliers of apparatus. At the same time it strengthened competition. Clauses 5 to 10 covered the regulatory framework based on a system of licences for all telecommunication operators, including British Telecom. Clause 8 was especially expanded to spell out in detail all the conditions to be included in licences issued to public operators in order to provide adequate safeguards for the customer. Most innovatory of all was the introduction of a scheme proposed by Stephen Littlechild (Professor of Commerce at the University of Birmingham) known as the RPI-X formula placing a ceiling on tariffs in the non-competitive areas which was subsequently embedded in British Telecom's licence. Two of the main changes were that BT would have to define its business to stop the cross-subsidisation of manufacturing activities by network activities, and any disruption of services would be subject to civil damages and not criminal proceedings.

These are the major practical differences between the first and second Telecommunication Bills, and just as the Bill changed its shape and detail during its stages in the Commons and the Lords, so did the Government's justification of the privatisation of British Telecom undergo subtle shifts. In July 1983, following the return of the Conservative Government, and the re-introduction of the Telecommunications Bill, a Department of Trade and Industry pamphlet, 'The Future of Telecommunications Government Policy Explained', put forward the rationale for British Telecom's privatisation. It placed the emphasis on competition and the free play of market forces and consumer choice. For these objectives to be met it was necessary to privatise British Telecom. The Bill now before the House would remove the obstacles to full competition and establish a common set of rules for all competitors. At the same time British Telecom would no longer be in the anomalous position of licensing its own competitors. Lord Cockfield, Chancellor of the Duchy of Lancaster, in moving the second reading of the Bill in the House of Lords, encapsulated the Government's motives in the following words:

> But true equality will only be achieved and full and fair competition attained if British Telecom ceases to be a nationalised industry and becomes an ordinary private sector company subject to the same rules, the same commercial disciplines, the same legal framework as its competitors and the generality of trade and industry.

Competition was the chosen solution to nationalisation which 'has not served the nation well either in terms of the efficient use of capital resources, or in terms of efficiency, good labour relations or good service to the customer'. Certainly the financial consequences of the sale were not unappealing since it would mean the Government could borrow less and British Telecom's demands for capital would be met by the private sector rather than the Government. Thus the desire to increase competition predominates over fiscal motives and this view is consistent with subsequent Government action in the rejection of a link up between British Telecom and IBM to provide value-added services and in the more recent decision to refer British Telecom's bid to buy Mitel, the Canadian manufacturer of telephone exchanges, to the Monopolies and Mergers Commission.

The committee stage of the Bill in the House of Commons had concentrated on ensuring adequate social provision and during the Bill's passage through the House of Lords there were comparatively few attacks on the general principle of privatisation. One major critic was Lord Weinstock, who attacked two fundamental assumptions of the Bill during its second reading. Firstly, he questioned whether privatisation by itself would generate efficiency: 'By itself, public ownership does not . . . emerge as an obstacle to efficiency . . . it is all a question of management', since British Telecom lies far behind France and Germany's state-owned industry in investment, revenue per employee and subscribers per employee. Secondly, he doubted that privatisation would create a competitive system since British Telecom would still remain the major network operator and the major provider of terminal equipment. In fact, BT would continue to dominate almost every aspect of Britain's telecommunication industry except manufacturing of equipment for which it is by far the biggest customer. Moreover, BT's extensive distribution channels and marketing muscle gave it a powerful advantage over potential competition. Lord Weinstock concluded that the legislation would 'place in private hands what has hitherto been a public service'. In spite of this last minute onslaught by Lord Weinstock on the fundamental premises of the Bill in January 1984 the Telecommunications Bill reached its final reading after 320 hours in Parliament and received the Royal Assent on 12 April 1984.

The passage of the Telecommunications Bills through Parliament was accompanied by behind-the-scene long-term discussions between British Telecom and the Department of Trade and Industry as well as Parliamentary lobbying by British Telecom. At issue were the form of British Telecom, the licence, the regulatory mechanism and British Telecom's financial structure—factors which would determine the success of the flotation and the future of British Telecom in the private sector.

With respect to the form of British Telecom two options were available—on the one hand, to sell BT as a single integrated organisation; on the other hand, to follow the example of the American Telephone and Telegraph Company which had been broken up into separate companies. The break-up of British Telecom was the option favoured by critics of the Bill who charged that if BT was sold as one corporate entity the Government was simply replacing a public monopoly with a private monopoly. Intensive lobbying by Sir George Jefferson, chairman of British Telecom, succeeded and four secretaries of state for industry accepted his economic and technological arguments to keep BT as a single entity. A break-up was judged impractical, not least because of the problem of finding buyers for British Telecom's unprofitable parts. It would also have diminished BT's ability to compete in world markets, where competition was conducted by multinationals. Furthermore, the UK needed a large company to face a liberalised home market and defend it against foreign competitors. The convergence of computing and telecommunications required companies with strong financial and technological resources, and since both sides agreed that BT's role in the future lay as the 'flagship' for Britain's information technology industry, keeping BT as one was the means to an end.

Naturally British Telecom had a strong interest in the provisions of the 25-year licence to be granted to them and the regulatory mechanism. The licence was the outcome of two years of consultation and unprecedented Parliamentary debate. BT's objective was to secure a 'light rein' of regulation and to resist more elaborate regulation which would result in it being perceived as a regulated utility, a poor basis for the forthcoming flotation. There was also the question of the exact way in which the regulatory mechanism would function. To which services should the price formula apply? How far should BT's permitted tariff increases be pegged below the rise in the Retail Price Index? The solution left British Telecom with the pricing flexibility needed to rebalance business and residential charges more closely to reflect costs, and with no price control on BT's international services or on apparatus supply, two growth areas of the future. A further question to be resolved was, how active would the newly created Office of Telecommunication be in using its powers? At the end of the long process Deputy Chairman, Deryk Vander Weyer, was able to assure prospective investors that 'We regard our regulatory environment as relatively mild and favourable to our growth'.

Another crucial area of discussion was BT's financial structure, the provisions for which formed Part V of the Telecommunications Bill. The outcome of this part of the Bill would determine whether British Telecom would appear as a pedestrian utility, attractive mainly for its dividend yield, or more like an exciting growth stock with a potential for capital appreciation. The latter perception required write-offs of long-term debt and flexibility in dealing with BT's pension liabilities, in order to present a 'clean' balance sheet for the prospectus.

The favourable outcomes in these crucial areas were not achieved by chance but by active, intensive long-term discussions behind the scenes with the Department of Industry. Simultaneously, British Telecom conducted a carefully constructed programme of briefings of politicians. For the first Bill BT's Corporate Affairs Department inaugurated a programme of meetings to establish a dialogue with MPs in order to explain their position and educate or reassure MPs on the future of services. Contact with MPs occurred at Party Conferences from 1982 and regular briefings to MPs were given at Westminster and BT's headquarters in Gresham Street. Throughout the winter of 1982–83 British Telecom maintained a presence during the House of Commons debates and through the Standing Committee (over 35 sittings). The ubiquitous presence of British Telecom in Parliament not only demonstrated the company's keen interest in the outcome but contributed to its ability to put up its own amendments! The first Bill was lost due to the election. The influx of nearly 150 new MPs elected in June 1983 and the re-introduction of the Bill necessitated a new round of briefings for MPs. The Corporate Affairs Department therefore arranged a programme over seven months whereby groups of new MPs were invited to visit BT for briefings. And again the company maintained a representative in the Commons to watch over the passage of the second Bill and even produced its own written Parliamentary briefings for MPs incorporating its views.

However, these were not the only or even the most important preparations which had to be made to groom British Telecom for its eventual launch upon the British public. Major internal structural changes had to be planned and put into execution even before the Bill had reached the statute book. The nature and significance of these changes will be the main theme of the next chapter. But before these can be usefully described it is necessary to say something about the character of British Telecom in 1983–84 and its earlier evolution as a major utility within the public sector.

Chapter 2

The Grooming of British Telecom

In 1983–84 British Telecom was rated as the fourth largest telephone authority in the world and was rapidly developing into a major information technology company. BT's core business was voice telephony for local, national and international services. It was also the major supplier of telephones, telecommunications equipment and a range of services and products including telex, radiophone, and radiopaging. British Telecom was increasingly developing new information and value-added services, and supplying office communication equipment and related services. In short it was already a long way from being just a telephone company and its size offered it the potential to dominate the emerging information markets. With turnover of £6.8 billion in 1983–84 and profit before tax of £990 million the figures already looked impressive before the challenge and the opportunities of privatisation appeared.

The take-off to this level of achievement was not the work of one decade or the product of any single new service or technology. It was the fruition of the long evolution of public communication services under the aegis of the old Post Office, from which British Telecom was not divorced until 1981.

Indeed, the potential significance of modern communications had long ago been grasped by Victorian Governments which had already had to address themselves to the fundamental question—should these services remain in private or public hands? It is an ironic comment on the revolution of 1984 that Victorian values of efficiency, economy and public welfare had dictated a large measure of State control. In 1879 because of public dissatisfaction with the service provided by private telegraph companies the Post Office obtained by statute the exclusive privilege of transmitting telegrams within the UK. The first telephone company in the UK was established in 1878, as The Telephone Company Ltd., using the patents of the inventor Alexander Graham Bell. In 1879, a rival, the Edison Telephone Company, was established but the two companies merged in 1880. The Government viewed the telephone operations by the United Telephone Company as an infringement of its telegraph monopoly and took action in the courts. Judgement was given for the Crown.

As a result, the Post Office acquired control over all telephone activity in Britain. It granted licences to companies in return for a 10 per cent royalty and set up its own telephone lines. The approach of the expiry dates of the main patents led to a series of amalgamations leaving the National Telephone Company with more than 73000 exchange lines in control of the entire telephone service with the exception of the 1233 subscribers served by the Post Office. The lack of competition and fiscal desire to obtain the revenues from trunk lines led to an agreement by the National Telephone Company to transfer the trunk lines to the Post Office in 1897. The National Telephone Company was then left to operate in London and the 213 exchange areas. Further expansion of the service encountered local authority opposition to the granting of way-leaves to the National Telephone Company for underground distribution of telephone lines. In 1898 at the recommendation of a select committee, the Government decided that the Postmaster General should grant licences to municipalities which applied for them, and that the Post Office should begin competing with the National Telephone Company at least in London. The Post Office began a metropolitan service on the strength of its own way-leave powers. In 1901 the Post Office agreed to construct the underground plant for distributing cables in London on behalf of its rival, and in return the National Telephone Company agreed to give free interconnection between the Post Office system and its own, thereby establishing co-operation rather than competition between the two systems. In contrast in the provinces there was an attempt towards genuine competition in the supply of the telephone service by municipalities and the National Telephone Company system. However, there was wasteful duplication of systems, there were no interconnection or standard engineering methods, and in areas of intense rivalry the National Telephone Company practised discriminatory pricing with the result that many municipal exchanges eventually passed into the hands of either the Post Office or the National Telephone Company. With the prospect of the end of the National Telephone Company's 31-year licence in 1911, agreement was reached in 1905 between the two suppliers for the National Telephone Company's system to be transferred to the Post Office. Thus at midnight on 31 December 1911 the systems of the National Telephone Company passed to the control of the Post Office with the exception of Hull where the municipal authority took over the Company's system and continues to operate it to this day. Hostility towards a private monopoly and the failure of the municipal experiment discredited the policy of competition and so in an age of *laissez-faire* the Post Office became the monopolist provider of the telephone service.*

---

* Hazlewood, A. (1953) 'The Origin of the State Telephone Service in Britain', *Oxford Economic Papers*, NS, and see Robinson, H. (1984) *The British Post Office, a History*.

As part of the old General Post Office British Telecom was until 1969 a Government department. In 1969 the Government changed the status of the Post Office from a Government department under the Postmaster General, to that of a public corporation under a chairman, appointed by and answerable to, the Secretary of State for Industry. This arrangement lasted until 1981 when British Telecom, having long enjoyed a monopoly in its basic task of supplying a telephone system and the telephones to attach to it, was given independence from the Post Office by the Telecommunications Act of 1981.

Logically, the first step towards privatisation following the 1984 Telecommunications Act was to transfer British Telecom's assets to the Government so that it would become the vendor of the share offer in November. Under Section 60, on 6 August 1984, the Government changed British Telecom's status from that of a public corporation to that of a public limited company and transferred the assets from the business into a wholly Government-owned public limited company under the Companies Acts.

There were two separate but related issues to be faced by the Government in the privatisation of British Telecom. The first centred upon the consequences of creating a private monopoly and the solutions were found in the creation of a regulatory environment featuring the creation of the Office of Telecommunications (OFTEL) headed by a Director General, and of the licence granted to British Telecom by the Secretary of State for Industry under section 7 of the Telecommunications Act 1984. The details of these will be reviewed below. The second problem was that of placing British Telecom on a commercial footing in readiness for its stock market debut. Three essential requirements stand out. A necessary prerequisite for the flotation was undoubtedly the development of an appropriate capital structure for British Telecom in order to reduce its gearing, an operation which required finding a middle course between presenting a balance sheet attractive to potential investors and avoiding criticisms of excessive writing-off of loans and liabilities. In addition the relationship between BT and the Government as major shareholder, with 49.8 per cent of the ordinary share capital, had to be spelt out to the investment community. Potential investors required an assurance of non-intervention in British Telecom's future commercial decisions. Finally, in the run up to the privatisation the Government had to install a credible management team with wide experience in the private sector to begin the long process of changing British Telecom's corporate culture from a civil service mentality to that of a profit-oriented market-led organisation which potential investors would view positively.

British Telecom's contribution was to respond to the new macro-environment. The changes were essentially managerial and included external recruitment of senior personnel, the introduction of new systems of accountability, decentralisation of decision-making, the establishment of independent profit centres, and most visible of all, a complete structural reorganisation of the company.

For the account that follows on the grooming of British Telecom I am heavily indebted to Jack Summerscale and Chris Wells of stockbrokers de Zoete and Bevan, authors of an omnibus in-depth survey designed to present British Telecom to the financial community.

THE CREATION OF A REGULATORY FRAMEWORK

Notwithstanding the element of competition introduced by the Telecommunications Act, the Government foresaw that British Telecom would continue to dominate the UK telecommunications industry for some years. The principal question before and after privatisation was (and still is) how to prevent British Telecom from using its dominant market position to exploit consumers. The Government's policy of liberalisation meant that British Telecom would no longer enjoy its exclusive privileges in any part of its business except in switched international services. And yet, given its enormous head-start and high market shares, effective competition was in reality light years away. BT possessed 80 per cent of the telephone and handset market, 90 per cent of the small private telephone exchange sector (PABX), and faced only one network competitor, Mercury. The idea of competition alone could not be relied upon to exert effective control or protection against monopoly practices, such as the ability to earn monopoly profits, tolerate inefficiencies, or use monopoly power to prevent, restrict or distort competition, both in old and in new markets. A regulatory system was therefore required to protect the consumer and to encourage the development of efficient and innovative telecommunications systems. But there had to be a balance. Over-regulation was just as incompatible as under-regulation. Over-regulation would jeopardise the net proceeds from the proposed offer for sale, since the success of the flotation and the revenue generated would depend inversely upon the perceived severity of the regulatory system. The degree of regulation would indirectly determine perceptions about British Telecom as a growth stock or utility and thereby in the long run affect the cost of British Telecom's capital-raising operations. As Professor Littlechild observed, 'If BT is perceived as a growth stock, its shares will stand at a higher price. The cost to BT of raising additional capital will thereby be reduced. If, on the other hand, BT is perceived as a reliable utility but one which is unlikely to grow, or likely to be constrained, then its shares will stand at a lower price. Raising additional capital will be more costly and this would raise the costs of providing telecommunications services in the UK'.* Under-regulation, on the other hand, would permit the perceived weaknesses of nationalised industries to continue—inefficiency, overmanning, disregard for consumers, and the ability to earn excess profits and pay uncompetitively high wages. To avoid such defects, the regulatory system

---

* Littlechild, S. C. (1983) *Regulation of British Telecommunications' Profitability*, Department of Trade and Industry.

would have several strands. It was designed to ensure that all reasonable demands for telecommunications services, in particular certain socially important services, are met, that operators are able to finance the provision of these services and that customers are provided with a variety of good quality services at reasonable prices. Clearly, in the philosophy which shaped the privatisation of BT regulation was regarded as a second-best to competition. Competition was envisaged as indisputably the most effective means of protecting consumers against monopoly power, but whilst effective competition was some years away, regulation might just prevent the worst excesses of monopoly and protect emerging competition. Thus, in addition to the existing legislation in the Fair Trading and Competition Acts, and the competition rules in Articles 85 and 86 of the Treaty of Rome, the Government created a regulatory system especially for telecommunications. This regulatory system has a number of features:

(a) the creation of the Office of Telecommunications headed by a Director General;
(b) the licence issued to BT under the provision of the Telecommunications Act 1984;
(c) the construction of an appropriate capital structure for BT;
(d) the defined relationship between the Government and British Telecom.

The Office of Telecommunications (OFTEL)

On 1 July 1984 the Office of Telecommunications was created with a politically independent Director General, appointed by the Government. The office was modelled on the Office of Fair Trading which over some years has set an often uncomfortable example of critical independence. The first Director General, Bryan Carsberg (a chartered accountant and professor of accounting at the London School of Economics) was appointed by the Secretary of State for a three-year period from July 1984. The Director General is supported by the Office of Telecommunications which has a staff of around 85. OFTEL acts as a watchdog supervising BT's compliance with the licence issued to it. However, OFTEL does not create policies for the telecommunications industry and may issue licences only with Government consent. Moreover, OFTEL does not make decisions, even on prices within the regulated area, but rather BT makes its pricing decisions and notifies OFTEL which checks to ensure that the decision is within the prescribed ceilings. In this role OFTEL is passive and it is hard to see a situation in which OFTEL could become 'proactive'—that is, initiating rather than reacting—given its limited resources.

Appointed by the Secretary of State for Industry the Director General operates within guidelines and criteria which appear impressive and include the following, defined by clause 3 of the 1984 Act:

1. To ensure British Telecom meets its statutory obligations. From observation or in response to complaints, it is up to OFTEL to ensure that the conditions in the licence which empowers BT to operate its network are complied with.
2. To ensure that all reasonable demands for telecommunications services throughout the UK are met. This requires the continued provision of the emergency services, public call box services, directory information, maritime services and services in rural areas.
3. To ensure British Telecom charges a 'fair' price for its services where possible, by encouraging competition in order to promote the interests of consumers and purchasers with regard to price, quality and variety of services offered.
4. To maintain and promote effective competition in telecommunications.*

If British Telecom is found to be within breach of its licence, and is recalcitrant OFTEL will be able to apply to the courts and BT may be liable for damages. It is also within the powers of the Director General to refer British Telecom to the Monopolies and Mergers Commission. OFTEL thus lies parallel to the liberalisation framework—a watchdog with borrowed teeth. But with no powers to quicken the pace of competition by creating new licences, it is difficult to envisage OFTEL attaining a 'proactive' role, and its impact on the environment may be limited.

The licence

Under the Post Office Acts and the British Telecommunications Act 1981 the Post Office, and subsequently BT, enjoyed an exemption from the requirement that all telecommunications suppliers be specifically licensed. This exemption continued until the 1984 Telecommunications Act which for the first time placed BT on an equal footing with its competitors in having to obtain and abide by licences for each of its operations. Today BT operate under a number of licences granted by the Secretary of State for Industry and the most important of these is the 25-year master licence to run its public telecommunications system (excluding Hull, which still has a separate system run by the local authority).

The conditions of the licence run to almost 100 A4 pages, but four aspects require elaboration. These are (a) the Littlechild formula for price control; (b) the terms of interconnection with other domestic networks; (c) the structural and financial separation of BT's equipment marketing and network activities; and (d) the definition of British Telecom's novel potential for diversification.

* The Telecommunications Act, 1984, Clause 3.

(a) Price control RPI−X

Before the 1984 Act, prices for British Telecom's services were set in the light of financial targets agreed with HM Government and following consultation with the Post Office User's National Council. Included in the licence issued to British Telecom (under s.7, Telecommunications Act 1984) was some control over the prices charged. There had been considerable debate in the early stages about the form such control should take. The initial view was that the control should be in the form of a return on capital, typical of that employed for telecommunications companies in the USA. The object of regulating BT's profitability until sufficient competition developed was solved by Professor Stephen Littlechild, in his influential report, 'Regulation of British Telecom's profitability', published in February 1983, three months after its commission by the Secretary of State for Industry. Perhaps the most innovatory part of the licence was the element of price control proposed by Littlechild and accepted by the Government. Control should lie not with profits but with tariffs. The licence requires British Telecom to ensure that certain prices do not rise faster than 3 percentage points less than the Retail Price Index. This RPI−3 formula (to be reviewed in 1989) covers a 'basket of services', namely the exchange line rental for business and residential subscribers and dialled calls originating and ending in the UK, i.e. local and trunk calls. Thus in any of the five years, beginning 1 August 1984, the weighted average increase in prices charged in British Telecom's traditional domestic services must be at least 3 percentage points below the RPI movement in each year. There is no price control on any element of international services as can be seen from Figure 2.1. The formula is ingenious for in contrast to America, the limit is on prices not on the rate of return. As a result there are strong incentives to improve profitability by becoming more efficient, rather than by merely increasing prices.

In addition to the control provided by the licence, BT has said that during the same five-year period it intends to hold increases in the rental for residential lines within a ceiling of 2 points per annum over the percentage increase in the

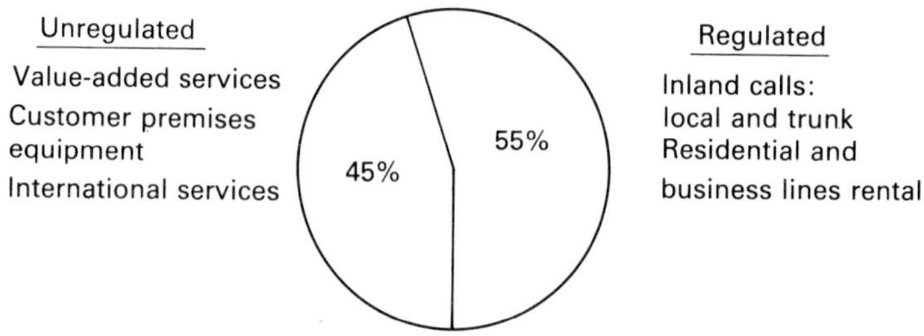

Figure 2.1 *Percentage of BT's current revenue subject to price regulation. Source: British Telecom.*

retail price index (RPI+2). Whilst British Telecom does not have to apply for an increase in prices, merely give notice to OFTEL, it will at the same time have to satisfy OFTEL, that it is meeting the RPI−3 formula. Subject to these controls and voluntary agreement, British Telecom can vary the annual increase within the individual basket of services. This flexibility allows British Telecom to use price as a marketing tool in meeting competition. It can adjust tariffs between local calls, trunk calls and line rentals in harmony with costs, and price its products competitively to reduce the scope for cream skimming by competitors. A start has been made, with the November 1984 increase in which charges for the unprofitable residential service rose more than on the lucrative long-distance traffic.

To summarise, the novel feature of British Telecom's regulatory environment is the RPI−3 formula. Whereas in the case of most privately owned utilities elsewhere in the world customers are protected from monopolistic exploitation by limiting the return on capital, in British Telecom's case this protection is applied by limiting allowable price increases to well below the rate of inflation. The Littlechild formula protects consumers, promotes efficient use of resources, provides incentives for management and provides for rising returns to potential shareholders.

(*b*) *Interconnection with other networks*

Under the terms of the 1984 Act and the licence British Telecom is obliged to permit the connection of its system to other systems, such as that of Hull and the Mercury systems which run under individual licences and those run under general licences such as private branch exchanges. The licence includes special arrangements for determining the terms on which interconnection is to be made between the British Telecom system and other individually licenced systems. However, in 1985 British Telecom and Mercury failed to agree on terms for the interconnection of their telephone networks. The matter went to the High Court, for the rules on interconnection are of critical importance to Mercury, a subsidiary of Cable and Wireless. Mercury is building a trunk telephone network consisting of an 800-mile optical fibre loop linking major English cities. Mercury has no local circuits of its own and the cost of installing them throughout the country would be huge. It therefore needs links with BT's network for most of its local connections, both for outgoing calls by Mercury customers and calls for Mercury customers who are more than local distance from its network. The matter at the present time lies unresolved, with considerable uncertainty as to the jurisdiction of OFTEL.

(*c*) *Separation of equipment and network activities*

The dominant position held by British Telecom in the provision of network services could prove a powerful weapon if used to promote its equipment sales and rental activities, hence damaging competitor's sales efforts. British Telecom will be required to establish separate accounting and reporting arrangements for the systems and apparatus supply business and has until 1 April 1987 to complete this process. The length of the transition phase reflects the massive exercise involved in changing from the system of nationalised industry accounting procedures and organisation.

(*d*) *Right of manufacture*

British Telecom has retained the right to manufacture, although any such activity will, after July 1986, have to be undertaken in a separate subsidiary of the organisation, so that separate accounts can be produced for network and manufacturing activities to demonstrate that there has been no cross-subsidisation. The licence also gives the Director General power to intervene where British Telecom unfairly cross-subsidises between its apparatus supply, its apparatus production business and its provision of mobile radio or value-added services. Furthermore British Telecom is not allowed to discriminate in the provision of its network service against subscribers using equipment purchased from a competitor.

THE REVISION OF BT's CAPITAL STRUCTURE

The next element of the preparatory stage was to determine the appropriate capital structure for British Telecom. Both the Government and British Telecom were anxious to see the company properly capitalised so that it had the financial strength needed not only to remain a major force within its own market but also to compete successfully overseas. In regard to the proposed privatisation of British Telecom, the Secretary of State announced in Summer 1984 the details of the restructuring of BT's balance sheet. These included:

1. A change in the arrangements relating to the £1250 million long-term liability in respect of the pension fund which was to become the responsibility of the statutory corporation and funded by the transfer of Government debentures in British Telecom.
2. The long-term debt owed to Government by the statutory corporation was restructured to become unsecured loan stock with a smooth maturity profile over a 25-year period and coupons closely related to market rates current at the time.
3. The creation of £750 million of redeemable preference shares carrying a gross interest rate of 11.75 per cent. These were to be non-voting and redeemable at British Telecom's discretion up to 35 years after issue.

The outcome of the changes to British Telecom's capital structure was to give

BT debt and preference capital which together amounted to some 75 per cent of its equity capital and an interest cover of around four times, somewhat stronger than most of the quoted United States telecommunication companies. These levels when combined with its expected strong cash flow enabled British Telecom to be perceived by many within the international financial community as a company of immense financial strength.

The legislation described above has set the path for a complete restructuring of not only British Telecom itself but the telecommunications service in the UK. Significant elements of competition have been introduced and success will depend upon the ability of the Secretary of State for Trade and Industry and in particular OFTEL to balance the demands of regulation and the opportunities of freedom.

THE RELATIONSHIP BETWEEN GOVERNMENT AND BRITISH TELECOM

After the offer for sale, the Government would continue to be the major shareholder accounting for 49.8 per cent of the issued ordinary shares. In addition a special share has the effect of giving the Secretary of State for Trade and Industry certain rights to prevent, in the public interest, excessive concentration of share-ownership. Notwithstanding the fact that the Government was to continue to be the major shareholder it was agreed ahead of the flotation that the Government would distance itself from the commercial operations of the company. The Government therefore gave a letter of non-interference in the normal affairs of the business, to ensure that British Telecom is not subject to short-term goals. That undertaking was set out in the prospectus as follows: 'HM Government does not intend to use its rights as an ordinary shareholder to intervene in the commercial decisions of British Telecom. It does not expect to vote its shareholdings on resolutions moved at General Meetings although it retains the power to do so.'

BRITISH TELECOM MANAGEMENT SHAKEUP

To cope initially with the liberalisation of the telecommunications market and subsequently with the change in British Telecom's status to a private sector company, the Government appointed a new team of board members with a wide range of outside business experience. The first significant appointment was that of Sir George Jefferson as chairman in October 1981. Sir George had been deputy chairman of the Post Office for about a year, but previously he had had a varied career in industry crowned by his appointment as Board Member of British Aerospace and chairman and chief executive of the Dynamics Group.

He was joined in 1981 by Deryk Vander Weyer, who became a part-time member of the board and was appointed deputy chairman in October 1983 after a long and distinguished career in banking, where he attained the chairmanship of Barclays Bank UK. The significance of these two appointments lies in the management philosophy they brought with them. Vander Weyer had established the first marketing department in a British bank in the late 1960s, and Sir George's declared philosophy is also market led; 'What business is about is giving a reliable, responsible service to every customer—big and small—at the right time in the right place at the right price.' Commitment to the marketing concept by the chairman and deputy chairman were essential prerequisites for a change in the company's culture, and these were reinforced by the appointment of John King in January 1984 as board member for marketing and corporate strategy after a career in IBM and Philips. Out of a main board of 13 (including two non-executive directors) nine had substantial outside business experience and the tier of senior management had also been strengthened by external recruitment, in order to change the corporate culture.

BRITISH TELECOM'S STRUCTURAL REORGANISATION

Perhaps the single most visible change was the structural reorganisation of the company into divisions and the decentralisation of decision making with the introduction of individual profit centres. The process began in 1982 when British Telecom was divided into four divisions: Inland, International, BT Enterprises, and Development and Procurement. But it was found that much of the manpower and most of the assets were concentrated in the Inland division. In 1983, therefore, the Inland division was split into a local and a national division, and the present structure can be seen in Figure 2.2.

Local Communications Services division (LCS)

Furthest from British Telecom's self-portrayal as a fast-moving high-technology business using satellites, microchips and optical fibres is the Local Communications Services division formed in 1983 when the Inland division was divided into a local and a national division in response to liberalisation and competition. LCS as the supplier of the local telephone network and handsets to some 20 million customers is the most familiar point of contact most people have with British Telecom. It is the largest operating division of British Telecom in terms of assets, revenues and manpower. In terms of revenue, LCS was by far the largest contributor to the group at around £4 billion or some 60 per cent of total sales in 1983. LCS handles over 84 million local telephone calls a day and provides the local access for over 14 million

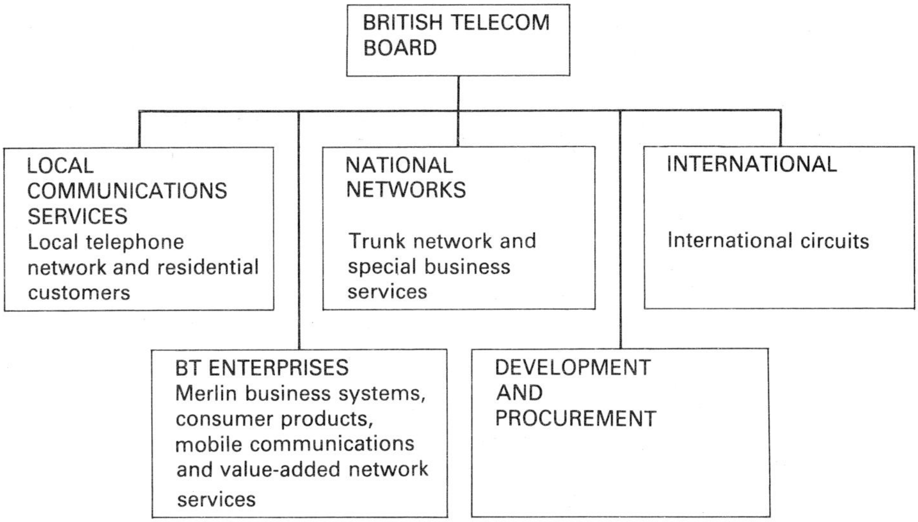

Figure 2.2 *British Telecom divisional structure 1984.*

trunk and nearly 600000 international calls daily. Running from the 6300 local exchanges are 16.6 million direct residential lines. In the business field there are 4.1 million lines from which are attached some 4 million extension phones. As regards private automatic branch exchanges (PABXs) there are some 65000 using 580000 lines and 3.5 million telephone extensions.

In effect LCS generates the bulk of British Telecom's revenues, although it then pays away to other divisions, National Networks, BT Enterprises and BT International the revenues appropriate to them. In the absence of the local telephone network the other divisions would have negligible revenues. The Public Switched Telephone Network (PSTN) business of LCS must be regarded as its core business. Not only is LCS the collection point for almost all National Networks revenues, it is the main marketing arm. To it devolves the responsibility for telephone call stimulation and most significant perhaps is the fact that the other divisions totally rely on LCS for the expansion of the network. The performance of LCS is therefore crucial to the group.

As regards assets, 80 per cent of the total are in the LCS division since LCS is the starting point for virtually the whole system. LCS employs nearly 200000 people, which represents around 83 per cent of the total BT workforce of whom over 50 per cent are engineers employed in maintaining the system. In addition to operating the local network LCS also sells, installs and maintains a wide range of products from handsets to short-haul private circuits and branch exchanges. LCS markets almost all of BT's equipment, most of which was customarily rented but is now beginning to be sold outright. It therefore is the main outlet for BT Enterprises, Merlin and consumer product equipment. In

Table 2.1 *British Telecom's main products*

	Market share %
Telephones	90
House Exchanges (Key systems)	95
Small PABXs (under 100 lines)	95
Large PABXs (over 100 lines)	30
Teleprinters	85
Answering machines and other telephone related equipment	30–40

Source: de Zóete & Bevan, British Telecom, June 1984

most markets it has a technical monopoly and therefore its market share remains very high. LCS also runs the UK operator services including directory enquiries, the emergency services reached by dialling 999 and telephone kiosks. LCS therefore links together many of the other divisional activities of British Telecom, firstly through its running of the local network as the first link in the chain for customers to call other local customers and to access the trunk and international networks; secondly there is a link through distribution of products and services for the entire company.

To move towards a more commercially oriented management structure the new division inherited an initial structure of 10 regions and 61 areas. However, to meet the board's goal of decentralisation of decision making, to balance senior management responsibilities, measure performance and introduce profit centres, LCS is being organised into territories: Scotland and Northern Ireland, North (Leeds), Central (Birmingham) and South (Bristol) already exist together with their 24 provincial districts. The new structure, it is felt, will give better control over the business, particularly with the appointment of high-calibre managers. As an example, the proposed reorganisation of British Telecom North-East is shown in Figure 2.3. The district is responsible for the marketing of all customer premises equipment, working with national networks account managers on large customer requirements. In addition it formulates the local policy on development, modernisation and utilisation of

Table 2.2 *Local Communications Services Division Functions*

The local telephone network (PSTN)
City business system leasing
Customer billing for all telephone usage (trunk, local and international)
Sale, rental and maintenance of all types of terminal equipment
Operator service
Emergency 999 service
Public call box telephones
Access and billing for X-Stream services run by National Networks

Source: de Zoete & Bevan, British Telecom, June 1984

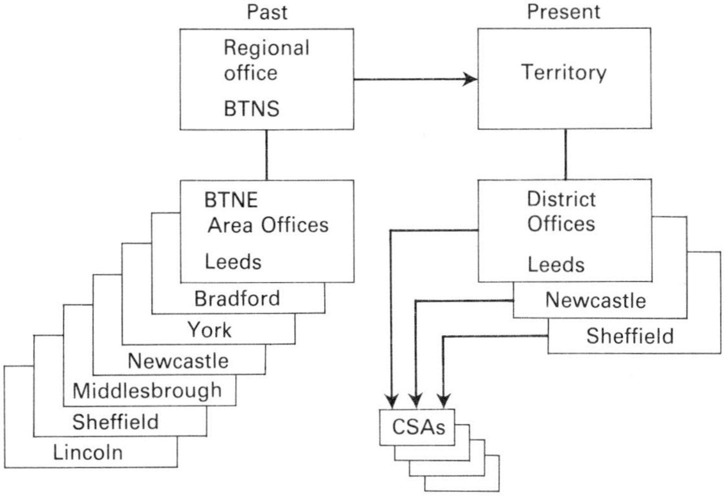

The new territory includes the former north-west RHQ and the north-west district offices of Manchester, Liverpool, and Preston/Blackburn

Figure 2.3 *BTNE organisational changes. Adapted from de Zoete & Bevan and British Telecom, June 1984.*

the network, and in conjunction with the relevant headquarters department it co-ordinates the local approach to new markets such as cable TV and office automation. Under the new structure there would be the danger that the local focus for customer service will be lost. As a result BT is establishing a large number of customer service areas (CSAs) of which there will be between 100 and 200.

LCS contributes just over half of BT's turnover and that is its strength. The weaknesses are many. Overheads and staffing levels are high. The main overhead burden other than depreciation is the wage bill, but much of the equipment is old for more than half of BT's local exchange equipment is the step-by-step electro-mechanical Strowger system. New technologies, in particular System X, have not been introduced in sufficient quantities to give rise to significant savings in labour maintenance costs. Only seven System X exchanges have been installed so far. Made by GEC and Plessey, they replace mechanical parts with microchips and provide routine new services such as redialling and rerouting calls automatically. They also permit revolutionary combinations of voice, data, text and video services in one line—a system which is now under trial in the City of London. For the future the application of microchip technology will offer substantial savings in costs and improvement of service, transforming a present weakness into a future strength and the potential new services on offer will counter the obvious threat of saturation on residential lines expected in the 1990s when there will be one telephone on average per household.

In both the short and long term, opportunities exist to increase revenues arising from the simple marketing tactic of increasing phone usage by advertising campaigns such as the highly succesful call-stimulation, 'It's-for-YOU-HOO' and 'Inphone' campaigns. British Telecom estimates that if every existing residential subscriber could be persuaded to make one additional trunk call a week profits would increase by £350 million whilst an additional local call could increase profits by a further £115 million a year. In the long run abundant opportunities should arise from the adoption of the marketing concept and the managerial revolution. The shift towards a market-led company culture has already begun, based on comprehensive consumer research based on Telcare, a computerised interview system imported from America to conduct 1.5 million customer telephone interviews a year, and to probe customer attitudes towards standards of provision, the repair service and billing system. This is no short-term inquiry but a long-term project established to monitor customer attitudes and complaints over time. A continuous effort to repair faults faster and eradicate the waiting list for phones are symptomatic of BT's new management philosophy, whilst the adoption of digital exchanges will reduce the incidence of crossed lines and poor connections, thus improving customer services and adding new ones. Similarly computerisation of manual records and streamlining of management systems will reduce costs. LCS is in the process of installing a computerised customer service system and a computerised management information system which is expected to be completed in 1987. The aim is that an on-line system will provide BT's staff with all relevant data on a customer at the time of the enquiry. Similarly, computerisation of the billing system and directory enquiries will contain costs, improve order-processing and fault-handling, and support a first-rate customer-responsive service.

A more market-oriented approach has been initiated likewise on the equipment side. Not only has the product range been strengthened and widened but there has been an increase in advertising in order to stimulate the market. Central direction to a major marketing thrust has come with the appointment to the board of John King, director for marketing and corporate strategy. His responsibility is the establishment of a centralised policy for marketing intelligence, marketing planning and pricing, products and local services, network marketing, advertising and sales promotion and field support.

The local network business has a monopoly which will hardly be touched in the next few years. However, with the ending of the 'prime instrument monopoly' (the jargon for the first telephone handset!) and the incursion of alternative suppliers the retail element of the telephone network is expanding. BT has begun to establish a High Street presence in order to sell its telephones and has extended its product range in order to tempt custom. Given its existing strong base and its new market-led philosophy it should maintain a high market share. Likewise, despite the regulatory limits on prices, the rate of return is not constrained and there are potentially very substantial margin improvements to be achieved from running the local network, provided BT can aggressively market its use and keep a tight rein on unit costs.

The National Networks Division (NN)

The onset of liberalisation brought about the split of the old Inland division into Local Communications Services and the National Networks division. It is the second largest operating division of British Telecom in terms of gross revenue and in isolation would rank as number 35 in the top 100 UK companies. NN is responsible for the long-distance trunk network which handles almost 4 billion trunk calls annually. The trunk network accounts for 90 per cent of the division's revenue and it is in this area that competition from Mercury is to be expected.

The business customer is the major user of the network, accounting for around 70 per cent of business. Moreover, a small number of customers account for a disproportionately high level of revenue. These are major business users such as Shell, GEC and Barclays Bank. To prevent defections to Mercury, and meet its business plan, British Telecom intends to invest over £1 billion by 1988 in order to provide a modernised fully digital streamlined trunk network. The existing public telephone network, an analogue system, was designed to carry speech. The sound waves created by the voice are converted into continuous and variable electrical waves which travel along the line and are reproduced at the receiver in their original form, together very often with a familiar crackling sound. With digital transmission, information is sent in coded form, offering speed and eradication of the customary interference on the line. Modern digital transmission systems use high technology micro-electronics and offer fast, high-quality transmission with improved reliability. The digital system can transmit speech, television, telegraph, facsimile, or data, thereby offering an expanded range of new services to business, particularly for office automation and electronic banking operations. The potential new services arising from a fully developed digital service include electronic mail, high-speed data transference, video text, electronic funds transfer, voice and data storage and audio and video conferencing facilities. In effect the new network will be multipurpose, high speed, intelligent, flexible and economical.

National Networks division also provides specialised services to the business sector through leasing of networks and lines for the transmission of voice or data. For example, large corporate customers can rent analogue private circuits to link their different sites all over the country. Alternatively, digital services, known as X-Stream, provides various forms of high speed data transmission. Lastly there is the provision of the Telex service with 100 000 customers connected. The trunk network business is highly profitable. British

Telecom faces only one competitor in this area until 1990—Mercury, a baby compared to the giant NN division, and Mercury is behind schedule for completing its figure-of-eight fibre optic cable system.

British Telecom International Division (BTI)

International communications is by far British Telecom's most profitable activity, producing profits of £320 million on a turnover of £1.2 billion in 1983–84. BTI has been the fastest growing segment of BT's business for many years though it is still much smaller than the major divisions LCS and NN. BTI defines its mission as being to provide, operate and maintain international and maritime services for the UK. International communications are carried out by a combination of satellite and submarine cables. BT International owns a major share in Intelsat, Inmarsat and Eutelsat, three satellite consortia which launch and run satellites and it owns 50 per cent of all the cables which land in the UK. BT International is one of the largest carriers of international traffic, accounting for almost 11 per cent of global satellite traffic and over 8 per cent of global cable traffic. By October 1984, BT International offered International Direct Dialling to 161 out of the 217 countries to which a telephone service is available. In the year ended 31 March 1984 telephone traffic to and from the UK amounted to 2012 million paid minutes. The bulk of international telephone traffic is generated from a limited number of routes and some 85 per cent of UK traffic is with 20 countries, whilst the largest single route, to the USA, accounts for 29 per cent (see Figure 2.4).

The international marketplace has always had some genuine competition because countries offer different rates for identical services. For example, it is more expensive for an American to telephone a friend in Britain than vice versa. Britain is one of the cheapest countries in the world for telecommunications; the cost of a three-minute call to America was some £1.62 in 1983 compared with the equivalent of £173 in 1927. Britain has won a substantial share of world communications traffic. Today, three fifths of American multinationals use Britain as their European communications centre. The UK is one of the world's principal transit centres, enabling traffic originating and terminating outside the UK to be routed via this country. BT International estimates that it handled more than 48 million paid minutes of transit traffic in the year ending 31 March 1984. Business customers account for 70 per cent of call revenues, the remainder originating from the residential sector calling friends and relatives, encouraged by an advertising campaign aimed at increasing consumer awareness of the relative cheapness of international phone calls.

BT International maintains an operator service for directory enquiries and other assistance to customers such as personal call facilities. In addition, for the

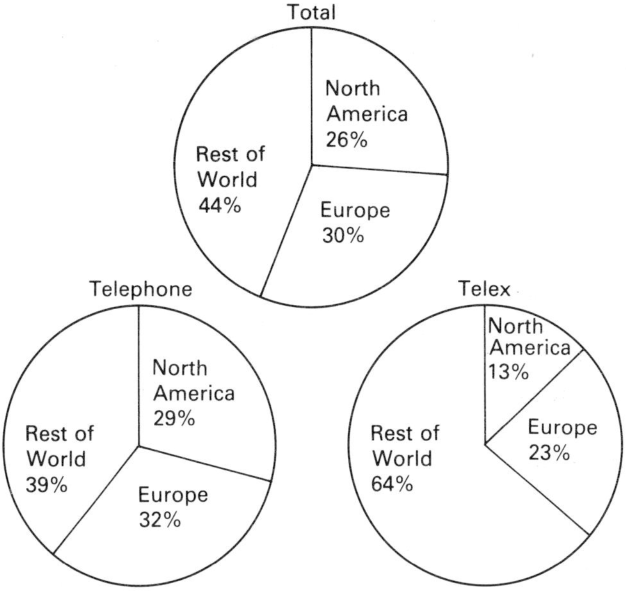

Figure 2.4 *Geographical split turnover. Source: de Zoete & Bevan, British Telecom, June 1984.*

international business customer BTI provides a telex service to 198 countries of which 186 can be reached by direct dialling. Telex provides 17–18 per cent of BTI's revenue. Whilst the final demise of telex has been forecast for some time by the introduction of alternative methods, such as facsimile, teletext, video text and electronic mail, which are all suitable channels for intra-firm communications across frontiers, telex is none the less still the universal channel for communicating with third parties and offers a major benefit—confirmation that the message has been delivered.

In addition to the public network, BTI provides private circuits for multinationals. The location of the UK and the price and service offered by BTI are found attractive, and sales teams visit the USA, Japan and Europe exhibiting and promoting this service. BTI dominates the Transatlantic leased circuit business. International leased circuits can link an office in the UK with another office in almost any country worldwide. Private leased circuits are economical where communications between two points exceed half an hour a day. Circuits for data, voice *and* data, or voice alone, are available and used by international business. BT International has nearly 60 per cent of all private circuits between Europe and North America, an achievement reflecting aggressive marketing and its lower charges compared to European competitors. It is in this highly profitable segment that Mercury, British Telecom's licensed competitor, plans to compete.

BT International offers a Datel service to 27 countries. This facilitates the transmision of data over the international public telephone network. Thus, special lines earmarked solely for the transmission of computer data are available for business enterprise. A recently introduced service is Bureaufax, where documents can be transmitted across the world in a few seconds. Operating in over 61 countries, delivery of such documents is by facsimile. Finally, mention should be made of teleconferencing and videoconferencing, the products of advanced technology. Teleconferencing can link up a maximum of 20 different groups worldwide. In 1984 Ford became the first company to use the European videoconferencing service of BTI business communications service. Every working day, Ford engineers and executives in Germany and England 'meet' for face-to-face talks without ever leaving their own plants. They are able to see and talk with colleagues, discuss pictures and graphics and evaluate vehicle components in detail. The development of telephones with pictures has been held back in the past because to transmit a normal television picture required the equivalent of 1000 telephone lines.

In 1983, in response to liberalisation, BT International was reorganised into seven separate units with five commercial executives: International Telephones, International Business Services, Operator Services, Maritime Radio Services and Telconsult (an international telecommunications consultancy) supported by International Lines and Satellite Systems executives. Each unit is headed by a chief executive reporting to the managing director. Each chief executive is responsible for his own financial contribution as a profit or cost centre.

With regard to marketing, the majority of British Telecom's UK customers do not have a direct relationship with BT International. The Inland divisions are the front line for promoting the use of BT International services, both in terms of increasing awareness of the international call related facilities of the telephone and telex networks and the other services offered by BT International such as leased circuits and telemessages. However, BTI does maintain direct contact with its major customers and customer organisations both at home and overseas. For example, all major business services such as leased lines, Bureaufax and Maritime Radio Services are marketed direct to major users. For business services, BTI organises seminars and exhibitions in the UK and overseas, and advertising is used to promote awareness of its services and their comparative cheapness.

BT International's revenues fall outside the RPI−3 formula since this Division is subject to the Government's objective of competition. Mercury is the only licensed carrier to compete with BT until the end of the decade and has chosen to enter the private leased circuit business, offering multinationals a parallel service to that of BT International. Leased lines is technically the easiest entry point into international communications but securing customers will be no easy matter. However, once Mercury has established an 'international gateway' with the establishment of a UK switching centre, the company can compete with BT in the switched telephony area. On both these fronts it is estimated that Mercury will obtain only a small market share during this decade, leaving BT International as the most profitable part of British Telecom.

British Telecom Enterprises (BTE)

British Telecom Enterprises was formed in November 1981 to enable British Telecom to focus upon newly emerging markets created by changes in customer needs, technology and the legal environment. BT Enterprises has a major role to play in expanding BT's involvement in supplying equipment to business and domestic consumers, in taking BT into new markets such as office automation and in spearheading BT's entry into new technologies such as cellular radio.

BT Enterprises is split into four sections, each with its own managing director who is also a member of the BTE board. The four sections and their principal missions are outlined in Table 2.3.

Table 2.3

Section	Equipment and service
Consumer products	Telephones, call makers, answering machines, key systems and small switches
Business Systems	Office products including PABXs, teleprinters and modems
Mobile communications	Paging and mobile radio systems
Value-added systems and services	Yellow Pages, Prestel, Spectrum services

The Consumer Products Section (CP)

The Consumer Products section develops, markets and supplies telephones and related equipment for the residential and business market, the latter accounting for some 63 per cent of the £180 million sales. It is the principal mission of CP to hold on to as much market share as possible since in January 1985 British Telecom lost its monopoly to supply the first telephone. The most important distribution channel open to CP is that of the telephone districts who have the facility and staff for marketing, order processing, installation and maintenance. The role of CP is therefore as a wholesaler to the telephone districts. It places bulk orders with suppliers such as GEC, Plessey and STC, and a small percentage with British Telecom's own factories, and then sells them on to the telephone districts, making a small turn on the transaction. Consumer Products' main customers therefore are BT telephone districts

(LCS), BT's Business Systems, other PBX suppliers, retail outlets serving the residential customer, such as Dixon's, Curry's and department stores, independent distributors and dealers serving the small business customer and individual customers gained from direct response advertising or via the newly established Phoneshops. At the moment LCS is the dominant outlet but retail outlets are increasing in importance. To hold market share as high as possible against new entrants to the market and maintain its position as the chief supplier to LCS, CP has embarked upon an aggressive product strategy. For example it has filled the large gaps in its product range with new, exciting and well designed handsets using a wider range of suppliers both at home and abroad. It has improved availability and delivery times and shifted towards a competitive pricing strategy. To ensure its competitive position and maintain its identity in the market place, CP insists on holding the intellectual property rights for all its distributed products. Its objective is to prevent substantial own-brand or private label production by outside suppliers who might seek to undercut British Telecom. At the same time, in order to shore up its position in a future highly competitive marketplace, BT has begun to use its own in-house research and development facilities to develop and manufacture its own telephones such as Slimtel and the Picture phone.

Business Systems (Merlin)

Merlin is concentrating on re-equipping the business sector with private automatic branch exchanges (PABXs) which are replacing the old office switchboards, a market worth some £200 million a year. Until July 1981, British Telecom had a monopoly to supply small private exchanges of less than 100 lines but was excluded from the large PABX market, that is private exchanges with more than 100 lines. Today, British Telecom still has a 95 per cent market share in the below 100 line market. In 1983 Merlin entered the large PABX market to compete with Plessey, Thorn, Ericsson, IBM, GEC and STC who were the approved manufacturers until the liberalisation in 1981. However, in general the larger and more important competitors are also Merlin's suppliers. For example, in the large PABX market, BT Enterprises Merlin is selling the Plessey-sourced Merlin DX against the identical product being sold by the Plessey sales force. Competition is therefore intense, with factors such as service and support playing a major role in customer choice, although pricing is still a major factor. Merlin, with its broad sales and maintenance capability through the LCS organisation together with its purchasing power has proved a strong competitor and now holds 40 per cent of the market for large PABXs which could be used as a platform to enter the fast-growing office automation and information technology sectors.

Mobile Communications (MC)

The Mobile Communications section of British Telecom Enterprises was formed in 1983 and operates the networks for radiopaging and mobile communications. The market for radiopaging and radiophone systems is open to competition, but British Telecom is the market leader in what is at the present time a relatively undeveloped market with only 4 per 1000 potential customers using radiopaging. The scope for growth of radiopaging is substantial as customers become aware of the benefits of being able to contact members of their workforce at any time, whilst at the same time British Telecom offers additional value-added services such as the alpha-numeric pager capable of receiving and displaying messages of up to 70 characters. In addition, a voice bank system has been set up so that the receiver of a paging call can retrieve voice messages left in a central electronic 'mailbox'.

British Telecom operates two radiophone networks, the old manual system in large conurbations due for closure in 1986 and the automatic direct dialling system recently extended beyond London. The third role of MC and one that is in its infancy is to establish a cellular radio network. Cellular radio offers the possibility of a completely portable telephone with full access to the telephone network and is based on new technology. The consortia for making cellular radios were licensed in 1982 by the Department of Trade and Industry—TSCR, which is jointly owned by British Telecom (51 per cent) and Securicor (49 per cent), and Racal Millicom. The two consortia will both build and operate networks which will interconnect with British Telecom's main network on the same commercial terms.

Value-added systems and services (VASS)

This section of BT Enterprises was also formed in December 1983 and concentrates on serving the business and mass consumer markets. A central role of this section is to expand the use of the telephone and the installed network. The section is responsible for two fairly well-known services, Yellow Pages and Prestel, and three less familiar services, Spectrum, National Data-Processing Services and New Information Services. Yellow Pages, launched in 1965, is the largest profit centre within the VASS section. There are 64 annual editions, distributed free to all businesses and homes with telephones throughout the UK. Every UK business with a telephone has a free entry under a classification of its choosing in its own area. As they are used by eight out of ten adults they have become a major advertising medium with around 250000 advertisers. Yellow Pages has about 50 per cent of the market for trade and classified directories. More recently, BT launched the Europages service, a business-to-business directory for importers and exporters in the UK, Holland, West Germany, France, Belgium and Italy.

Technologically the most adventurous segment of VASS is Prestel, launched in 1979 as the world's first public viewdata service. It links adapted television screens to computers through ordinary telephone lines so information can be delivered directly into homes and offices. Customer acceptance has been slow, but is now picking up at the rate of around 2000 subscribers a month to a total of around 60 000, almost certainly because of the introduction of interactive services such as Homelink for home banking and Micronet 800 to link home computer users. There remains considerable scope for mass home-banking, home-shopping and computer services. In the business sector specialised services have been developed for particular market segments. For example, Prestel Travel Service is used by 90 per cent of travel agents and Skytrack offers an on-line airline reservation system whilst Prestel Citiservice provides rapidly up-dated information on stocks, commodities and foreign exchange mostly for users located outside the City.

Spectrum is a form of value-added service provided over the public switched telephone network. Formed in 1981, Spectrum offers a broad range of services including the following:

(a) Telecom Gold—an electronic mail-box service;
(b) Telecom Red—a centralised alarm system against intruders and fire;
(c) Telecom Tan—a computer-aided telephone answering service;
(d) Telemarketing—a marketing and questionnaire service via the telephone;
(e) Telecom Audioconferencing—offers a three-way or more link through BT that can be booked for conference calls.

The National Data-Processing Service is a mature and profitable data processing business which handles 90 per cent of all air freight processing and 70 per cent of all customs import entry pricing. New Information Service is a new business, set up to explore and exploit the opportunities created by the establishment of new, and often competing networks. The largest business in this group is Cable Interactive Services, providing a consultancy in interactive services to British Telecom Cable and packages of services to home banking, shopping and information retrieval operators. The value-added systems and services section is currently growing very rapidly, providing a turnover exceeding £100 million and enjoying a growth rate of nearly 40 per cent.

BT Enterprises therefore has been formed to market newly liberalised subscriber equipment and services. It will compete both against its three main traditional suppliers, GEC, Plessey and STC and against such international giants as IBM and Xerox.

Development and Procurement Division

Development and Procurement is the fifth division of British Telecom and is responsible for research as well as development and procurement. British Telecom spent just under £200 million in 1984–85 on research and development in order to keep British Telecom's operations abreast of the latest most cost-effective technology. The biggest challenge has been the development of System X, designed by British Telecom and made by Plessey and GEC, to revolutionise the exchange network by replacing the old analogue exchanges with digital System X exchanges. The key strategy is to introduce digital services to the business segment as quickly as possible and then apply it to the rest of the network. Likewise the application of research to modernise the transmission network is underway. The old analogue transmission network is being replaced with newer digital systems including fibre optic links, not only to meet the need for re-equipping the network, but also to add substantial capacity to meet increased demand for new services.

The role of procurement is essentially to provide the operating divisions with the equipment to service customers efficiently. The Procurement Executive is responsible for the purchase of all equipment for use in the telephone network. As its name suggests, the department undertakes equipment selection and rate of purchase decisions, and sets standards in consultation with the operating divisions concerned, but when the initial policy has been laid down it is up to the Procurement Executive to ensure that the necessary equipment arrives on time at the required price and quality.

The advent of competition and changing technology has had a far reaching impact on the Procurement Executive. There is now an increased emphasis on value for money or cost-effectiveness for all purchasing activity. The advances in new technology have shortened product life-cycles and increased the need for a broader range of suppliers both at home and abroad. The research and development work, mainly conducted at Martlesham, and the work of the Procurement Executive are crucial to the operation and profitability of British Telecom. They are essential preconditions for its continuing competitiveness and responsiveness to customer needs.

The above account has focused on the managerial and organisation developments undertaken to prepare British Telecom for its subsequent flotation to the traditional investment community. However, to reach a wider and less experienced audience, hitherto outside the Stock Market, there had to be a more colourful, inventive and exciting approach to the marketing of British Telecom and it is the mounting of this campaign which is the theme of the following chapters.

Chapter 3

The UK Retail Market

THE DRAMATIS PERSONAE

As the sponsoring department the Department of Trade and Industry was intimately involved in every stage of the planning of the British Telecom flotation. The Telecommunications Act 1984 vested in the Secretary of State sundry powers and responsibilities. He was endowed with the entire shares and debentures of British Telecom and it was from his hands that the markets bought three billion shares. He was given the power to transfer all property, rights and liabilities in British Telecom to a successor company chosen by him on a transfer day selected by him as 6 August 1984. The Secretary of State was also the ultimate licensing authority for telecommunications activities after consultation with the Director General of OFTEL.

But the Department's responsibilities extended well beyond these formal powers to the more complex problems of marketing the BT shares. DTI officials were represented on all the major committees which planned the financial terms, the prospectus, the marketing and logistics of the sale. The Department made the crucial decisions on key advisers, the choice of Kleinwort Benson as the issuing house, and of Hoare Govett, Scrimgeour, Kemp-Gee, and De Zoete and Bevan as three of the four stockbrokers to the issue. The Department and BT in conjunction agreed on the appointment of Dewe Rogerson and Dorland to handle the public relations, marketing and advertising of the BT share offer. It was the Department who ultimately authorised the necessary expenditure involved in designing and marketing the offer for sale. The Department of Trade and Industry officials were therefore essential members of a team, and it is inconceivable that any significant decisions were taken without their approval.

Following a competitive presentation by five leading merchant banks, Kleinwort Benson was appointed as financial adviser to HM Government in August 1982. Its primary role in connection with the flotation of British Telecom was to advise the Government on the appropriate capital structure for

British Telecom, on the impact of the proposed regulatory regime, on market capacity for the sale of shares and on the structuring and marketing of the offer. More generally, however, Kleinwort Benson acted as the co-ordinator of all the activities of a very wide range of professional advisers and others involved in the flotation and as the focal point for advice to the Government. Their role ranged from the preparation of the prospectus (in conjunction with British Telecom, BT's financial advisers, HM Government and lawyers both in the UK and overseas) to participation in the retail marketing campaign conducted by the Government's public relations adviser and advertising agency and by a network of 22 regional stockbrokers in the United Kingdom. Kleinwort Benson was responsible for the arrangements in the UK to print more than six million prospectuses with associated documentation and to process more than two million applications. The issuing house was also responsible for overseas offers in Canada, Japan, Switzerland and the United States which were arranged on behalf of the Government by overseas investment banks. The team was led by Martin Jacomb, Lord Rockley and David Clementi, the director with day-to-day responsibility.

S. G. Warburg acted as financial advisers to British Telecom on all aspects of the privatisation, having previously advised the company on other matters. Whilst working together with the other advisers towards the success of the flotation, Warburg's main function was to offer independent advice to British Telecom on areas in which there was or could have been a divergence of interests between British Telecom and the Government in order to achieve a result which was in the best interests of both the company and its prospective shareholders. These areas included certain aspects of the legal and regulatory framework in which British Telecom would operate, the determination of the capital structure, the restrictions on future sales of shares and loan stock by the Government after the flotation, the fixing of the dividend payable, the marketing of the shares, and the method of sale at home and abroad. Warburg also jointly reviewed the profit and working capital forecasts of British Telecom. The team was led by Michael Valentine, a senior director, and Nicholas Fry, an executive director of the bank.

Dewe Rogerson, an advertising, marketing and leading financial public relations agency, was appointed in December 1983 jointly by the Department of Trade and Industry and British Telecom to develop and execute the marketing and communications campaign for the British Telecom share offer. The campaign was to comprise practically all the main communication techniques, including public relations, design, literature, mailings and presentations culminating in large-scale publicity for the prospectus. Dewe Rogerson was responsible for advising on, and guiding on a day-to-day basis, the overall marketing campaign. As an executive arm of the flotation marketing sub-committee, Dewe Rogerson was responsible for initiating most of the proposals considered by the committee and, after approval, arranging for their execution.

For example, much of the literature, video and films for the BT share offer made available to the public and financial intermediaries was conceived and designed by Dewe Rogerson who then arranged for their production and distribution subject to the control of the flotation marketing sub-committee. The idea of the British Telecom Share Information Office and the associated enquiry handling service originated with, and was largely put into practice by, Dewe Rogerson. The team was led by Roddy Dewe and Anthony Carlisle.

Dorland Advertising, one of Britain's largest advertising agencies, was appointed to create and run the advertising campaign that proved to be one of the principal motivating factors in persuading private, inexperienced investors to invest in British Telecom. Dorland as BT's existing corporate advertising agency had already developed the highly visible 'Power Behind the Button' campaign, which since November 1983 had been significantly changing the public's attitudes and perceptions of British Telecom. In the campaign for British Telecom's flotation, Dorland applied the disciplines of creative consumer advertising, familiar to its existing clientele but very new territory for those involved in the financial world. Dorland's ability to negotiate cost-efficient television airtime and press space was already well known to British Telecom. Their skill was demonstrated for the flotation in a negotiation that permitted maximum flexibility in the weight of advertising as the campaign progressed. This allowed the tempo of the campaign to be regulated to the needs of potential investors using the information supplied by regular consumer research. The team for the flotation was headed by James Hudson, creative work was under the direction of Royston Taylor, and Nick Farley had responsibility for media. The 100-strong agency team was supervised by Dorland chairman Jack Rubins.

Market & Opinion Research International (MORI) had for several years been conducting tracking studies and other research on the image of British Telecom. Thus when the flotation team was formed MORI were asked to present their results to the flotation marketing sub-committee chaired by British Telecom. MORI were appointed to conduct quantitative and qualitative research into the attitudes of the various segments important to the success of the flotation, including the general public, the City, the business and financial press and 'advisers' including stockbrokers, bank managers, solicitors and accountants. As a guide to the research programme and the influences on share buying the campaign team employed a conceptual model, based on an original concept first developed for the BT flotation by MORI. In this instance qualitative research was undertaken using group discussions with likely share buyers to see what part various influences might play in the determination of whether or not to buy shares and if so how many. The presentations to the flotation marketing sub-committee were led by Bob Worcester and the research executed by Cary Martin.

OBJECTIVES

For the Government, as the vendor of British Telecom, and Kleinwort Benson, as the issuing house, there were three predominant and interrelated objectives for the privatisation of British Telecom.

The first natural objective for both parties was a full subscription to the £3.9 billion offer for sale at an optimal price. For the Government, a successful sale of 51 per cent of British Telecom would allow the Treasury to report that it was keeping to its public expenditure plans and containing the public sector borrowing requirement. In addition it was very important to avoid a situation where a large proportion of the issue was left with the underwriters, which could jeopardise the whole programme of future asset sales. For Kleinwort Benson, in common with any issuing house, selling all the shares at a fair price and clearing the market was a measure of professional success and therefore the primary objective.

Secondly, there was the wider share-ownership objective, based partly on ideological commitment and partly on fiscal necessity. John Moore, a Treasury minister, speaking to the Wider Share Ownership Council Forum in October 1984, expressed the Conservative Government's ambitious philosophy: 'Our aim is to establish a people's capital market, to bring capitalism to the place of work, to the High Street and even the home.' But undoubtedly of equal importance was the problem of containing a public sector borrowing requirement which was threatening to exceed its limits, and to do so by achieving a flotation which in sheer scale was some six times larger than any previous UK issue and four times larger than any other capital-raising operation in the world. The very size of such an issue of 3012 million shares itself created the practical necessity of extending the funding burden beyond the confines of the traditional capital market. Thus, there was a convergence of political ideology and financial necessity in inducing an unprecedented and ambitious retail marketing solution.

Thirdly, it was recognised that if this issue were to be successful, it needed to appeal to as wide an audience as possible, not merely from the domestic market but from investors abroad. Indeed if British Telecom was to play an important role overseas in telecommunications and information technology it needed to have the support of international investors. A simultaneous offering of shares abroad would spread the funding burden and automatically give British Telecom an international company status.

These three interdependent objectives necessarily determined that there would be three markets to be targeted in the sale of British Telecom shares.

THE THREE MARKETS

The first of these was the traditional market for new issues and customary underwriters of offers for sale, the UK financial institutions. This traditional sector ultimately absorbed 47 per cent of the share issue. Second, following the precedent set by BP and Reuters, there was the overseas counterpart of the UK institutional sector, which was limited to 14 per cent of the stock. The third one, and the unique one, was the British public—the unpredictable and hitherto largely untapped 'retail market'. This was the challenge: to identify, and communicate with a vast audience which was defined initially as all adults with savings over £2000 and in particular the 16 million or so residential subscribers to British Telecom's telephone network. Not only did it make sense to match the size of the offer with the widest possible potential audience, but it was a distinct advantage to British Telecom to secure as many as possible of its customers as shareholders. Not only would this mean that they have a wider understanding of what the company does, it would also ensure that the company maintained its new-found attitude of responsiveness to the marketplace and consumers. In the event 39 per cent of the shares were allocated to the public among which British Telecom employees formed what might be called a fourth, though less significant, market. Indeed, the Government was anxious to get the maximum support from British Telecom's employees for the transformation of the company into the private sector domain. The official union position during the run-up was one of hostility to the privatisation move. However, direct communication between the board and all employees, a full employee communications programme and the Government-funded employee share schemes succeeded in winning over 95 per cent of the workforce. The Government offered employees three schemes. Firstly, each employee on application was given £70 worth of shares free. Secondly, each employee received two shares for every one bought subject to a maximum of £300. Thirdly, employees were given preferential allotment in the public offer. In the event it was clear that few employees heeded the official view of their union. Some 96 per cent of British Telecom's workforce applied for shares and accounted for 5 per cent of the total number of shares on sale.

The three distinct markets thus defined became the focus of the marketing campaigns to sell British Telecom shares. Fundamental to the success of any issue is a situation where demand for the shares exceeds the supply of shares, but given the scale of the British Telecom share issue, this ideal seemed at first an impossible one. Everyone knew there were over 3 billion shares for sale—a promise of super-abundance. The solution lay in changing the markets' perceptions. Each of the three target audiences understood through marketing and public relations activity that there were other higher sources of demand: thus Kleinwort Benson played off established overseas interest in British Telecom shares against UK institutional interest. But the catalyst in this was the retail market. If the terms of the offer were sufficiently attractive and marketing communications could generate a meaningful investor demand from the public then the overall funding burden of the financial institutions could be reduced. However, if the public thought the offer was attractive to them then it was likely that the offer would be attractive to the financial institutions. It was this interaction between the retail and institutional sectors that was to prove vital to the outcome of the issue. The snowballing of public interest, identified by MORI surveys and fed by press analyses, stimulated the perceptions of fund managers and financial advisers that British Telecom shares would prove a good buy.

THE FLOTATION MARKETING SUB COMMITTEE

An unusual team was created in July 1983 to prepare for the marketing of the British Telecom offer for sale. The Flotation Marketing Sub Committee was initially chaired by Peter Young, British Telecom's director of Corporate Relations, and from mid-1984 by John King, the new British Telecom board member for marketing and corporate strategy.* The team included senior executives of British Telecom, representatives of the two merchant banks—Kleinwort Benson acting for the Government and S. G. Warburg for British Telecom, stockbrokers Hoare Govett and Cazenove, civil servants from the Department of Trade and Industry and the Treasury, Dewe Rogerson as public relations advisers, Dorland Advertising and in some instances lawyers for British Telecom and the Government. From November 1983 this committee of some 20–25 people met fortnightly at British Telecom's headquarters in Gresham Street, occasionally sitting from 4 p.m. until late into the night. The findings from MORI surveys were previewed at a preliminary meeting held at 3.30 p.m. before the start of each formal session, and intermediate meetings on an *ad hoc* basis were called when necessary to review publicity material and advertising before it was broadcast.

The marketing committee was in fact the ideal solution to the problem of co-ordinating the marketing programmes for a gigantic new issue. As in any new issue the key to success would lie in the control and timing of the release of information. Equally, there would be thorny problems posed by statutory constraints and by the legal requirements of the prospectus document. For each and every statement in the prospectus the Secretary of State and BT directors were primarily accountable and lawyers must verify all claims. Anything that is said in marketing and promotion must accord with the prospectus: they must not conflict. This therefore requires the co-ordination of bankers, accountants, lawyers and marketeers. The division between ownership and management of British Telecom itself required a forum for policy decision making, discussion of marketing strategy and allocation of marketing costs between British Telecom and the Government. The logic behind this last requirement was, that whilst the Government was willing to pay for the larger

---

* Hereforward this is referred to as the marketing committee. Formerly it was a sub committee of the prospectus committee.

part of marketing specifically directed at the flotation, part of the promotion would be of direct benefit to British Telecom as a company. Finally, the envisaged scale of a marketing campaign designed to sell almost £4 billion worth of shares itself required co-ordination, direction and management. There were no rules or precedents to follow, and combined with this commitment to an innovatory approach in terms of scale, techniques, complexity and length, it was essential for all parties to the marketing of the issue to have a forum for discussion, to make policy decisions, to approve expenditure, to monitor progress and to measure results.

In the course of its life the marketing committee approved a large-scale integrated communications programme of advertising and public relations, a marketing package specifically aimed at financial intermediaries, exhibitions and visits, the provision of supporting literature through the largest direct mail programme in the world, and an intensive market research programme to check, monitor and forecast public demand. In effect the marketing committee approved, budgeted and monitored a host of marketing strategies as can be seen in Figure 3.1 for the DTI's offer campaign. All advertisements and flotation material had to be checked for propriety, visual appeal and effectiveness whilst media schedules were scrutinised for cost-effectiveness and penetration. To underpin the educational and informational emphasis of the marketing drive, numerous items were approved for distribution to the public and financial intermediaries during the course of the campaign. There were 24 major pieces of print, with print runs from 100000 to 20 million, including 1.5 million British Telecom corporate brochures, 20 million telephone-bill leaflets, 100000 brochures showing British Telecom products and services, a million fact sheets and 1.4 million Stock Exchange booklets all bearing a statement that they had been: 'Published and distributed by Kleinwort Benson Limited, on behalf of HM Government'. Banks and Post Offices also carried display material including leaflets and posters. Three main videos were produced as a back-up for the PLC train tour of 17 cities, and after the publication of the Pathfinder prospectus on 26 October there were roadshows in the UK and overseas. In addition a video made specifically in association with Barclays Bank, featuring Noel Edmonds, was made available to clearing banks to help them deal with the public's response. Altogether 42.5 million units of print and film were distributed. Furthermore, the British Telecom Share Information Office (BTSIO), the biggest telephone answering service in the world, which handled more than 1.3 million public enquiries, not only reflected an innovatory decision but represented a sensitive measure of advertising's quantitative impact. Another innovation was the decision to produce more than six million eight-page 'mini' prospectuses for small investors which were sent to over a million homes and widely distributed through banks, post offices and financial intermediaries in addition to the full prospectus published in four national newspapers. The full 64-page document was considered likely to put off many

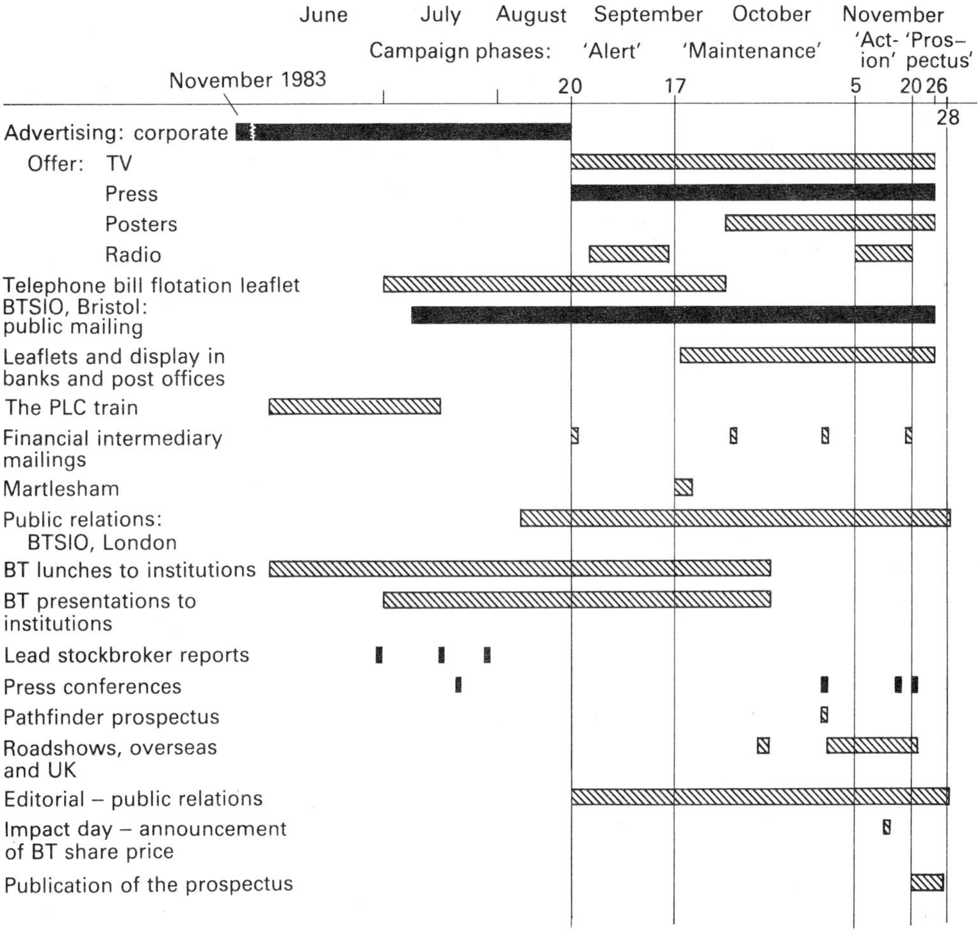

Figure 3.1 *Timetable of major communications, June–November 1984 (corporate advertising began in November 1983).*

first time share buyers, so the Government introduced an amendment to the Telecommunications Bill exempting the 'mini' prospectus from the requirements of the Companies Act, 1948 and the Prevention of Fraud (Investments) Act 1958. As it happens, as part of the process of adopting the community's directives it has since become possible for new issues generally to have a full *and* abridged or 'mini' prospectus providing Stock Exchange rules are met.

Each of these separate yet integrated programmes required examination, justification, auditing, approval and a budget. Each meeting assessed progress reports or briefings as to the current state of play and planning for the future since lead times for ordering paper, arranging design, printing and distribution of large quantities of literature were considerable. From conception to completion required time and continuous monitoring.

Every marketing strategy proposed to the committee by Dewe Rogerson was examined in terms of specific tasks and objectives and the estimated cost scrutinised prior to a search for cheaper alternatives. The drive for value for money combined with the flexibility of techniques available meant that considerable attention was paid to strategic planning, detail and budgets. Unlike a company's budget setting where an aggregate amount is then divided between advertising and below-the-line promotion, the marketing committee approved individual programmes on the basis of best estimates of expenditure linked to stated assumptions on numbers likely to be required. In most cases many of the costs were demand-led, in the sense that the quantities of informatory literature printed and distributed were directly dependent upon the scale of the public's response. For example, whilst the cost of enquiry handling was estimated in May, no one predicted the increased expenditure required from the additional 25 telephone lines added in the fortnight before the publication of the prospectus in order to cope with the demand. Similarly, the number of 'mini' prospectuses and the full prospectus required to be printed were difficult to gauge at the outset since they too were demand-led and the difference in cost estimates and actual expenditure could be considerable given the variety of programmes and the large numbers involved.

MARKET RESEARCH

The backbone of the retail marketing campaign lay in market research conducted by Market & Opinion Research International (MORI) commissioned by Dewe Rogerson on behalf of British Telecom and the Department of Trade and Industry. Market research provided a continuous flow of information on the public's attitude towards British Telecom, towards investment and financial products in general, and towards particular features of the offer for sale such as the shareholder benefits and intended investment levels. Qualitative research and quantitative tracking studies of attitudes over many months not only provided essential guidance to the communication strategy but also acted as a monitor for the effectiveness of the communication strategy as a whole. Equally important were the continuous and *ad-hoc* surveys for the construction of the British Telecom decision model used to forecast the outcome of the offer for sale, particularly as to the number of applications and the uptake of the bonus issue *vis-à-vis* the telephone voucher benefit.

Corporate image tracking studies

Already in Autumn 1980, in anticipation of liberalisation of telecommunications, British Telecom had commissioned MORI to begin long-term survey research in the form of quantitative monthly tracking studies of British

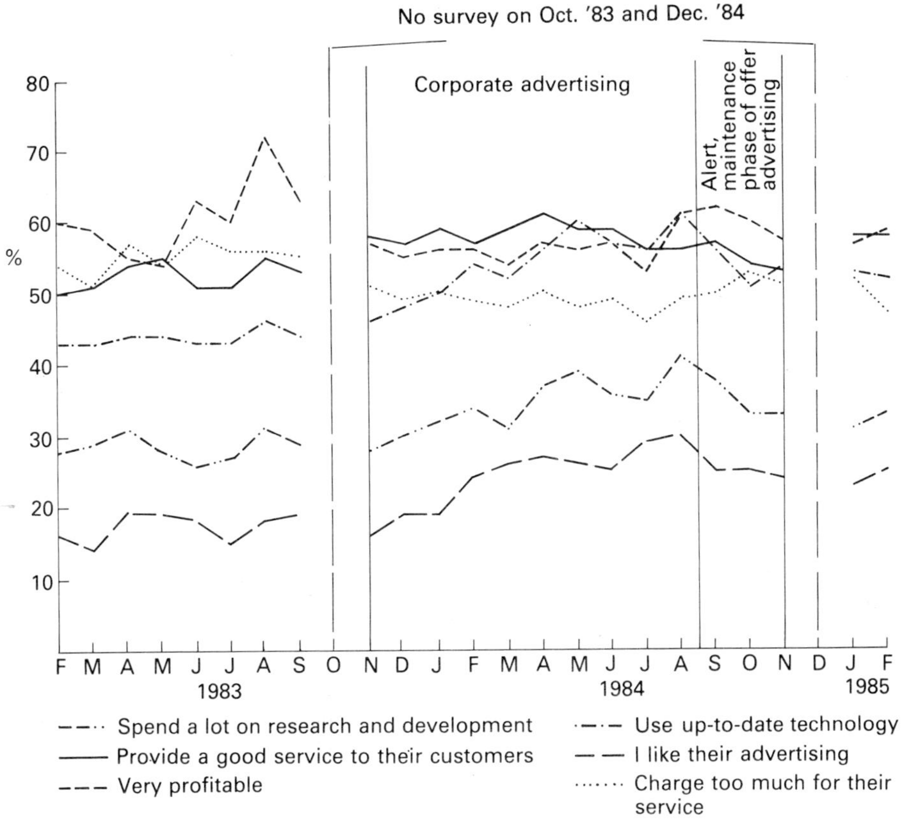

Figure 3.2 *Corporate image of British Telecom among the general public. Source: MORI corporate image monthly tracking studies.*

Telecom's corporate image amongst the general public the results of which are illustrated in Figure 3.2. An organisation's image is the result of all experiences, impressions, beliefs, feelings and knowledge people have of it. All companies project an image, whether it is deliberately conceived and cultivated or left to chance. But some such image is crucial. It can shape people's behaviour towards a company, influence their predispositions to try its products or services, pay attention to and believe its communications, speak favourably of it to others and listen to its point of view in controversies. The dimensions that make up a company's image are numerous but there are five major elements. Firstly there is the firm's civic role, measured in terms of its contribution to the quality of life and importance to the community as well as its perceived social responsiveness. Secondly there is its reputation and customer relations, measured by product quality, reliability of service, fairness in prices charged and the degree to which the firm lives up to its guarantees and promises. Thirdly there is the quality of its corporate leadership measured by such factors as perceptions of future prospects as a good company to own shares in,

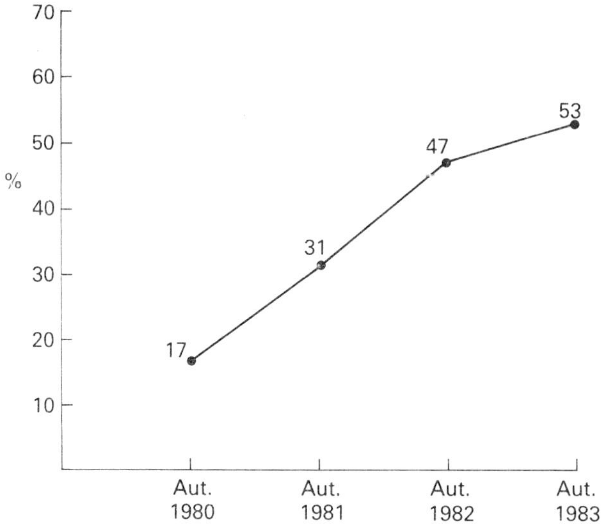

Figure 3.3 *Familiarity with British Telecom. Source: MORI.*

being modern, efficient, technically go-ahead, profitable and well-managed. Fourthly, there is the employer role, measured by the perceived state of the firm's industrial relations and staffing levels. Finally there is the communications dimension, measured by its advertising, its responsiveness to public suggestions, the information emanating from the company and the favourability of media comment.

British Telecom's image strengths in the Autumn of 1983 were considerable. It was seen in a positive light in its contribution to improving the quality of life, in the reliability of its service and importance to the community, in the state of its technical advancement and in its profitability. The five key dimensions interact to produce a general corporate image which can be tapped by two measures, familiarity and favourability. The familiarity measure represents a combination of objective factors such as experience and knowledge and subjective ones such as impressions and feelings. There is a positive correlation between how well a company is known and how favourably it is regarded. As can be seen from Figures 3.3 and 3.4 both measures, familiarity and favourability of British Telecom amongst the general public, had climbed gradually in the three years from Autumn 1980. In Autumn 1983, a year before the company's flotation but at a time when the marketing strategy was beginning to be seriously considered, British Telecom was in a happy position of enjoying considerable public support. Already 69 per cent of the public were favourably disposed towards the company and 53 per cent felt they knew the company well, given the ubiquitousness of its telephone service. Favourability was disproportionately higher than familiarity with British Telecom, and indeed, compared with other companies (Figure 3.5) British Telecom's position is among the most highly thought-of British companies, exceeded only by

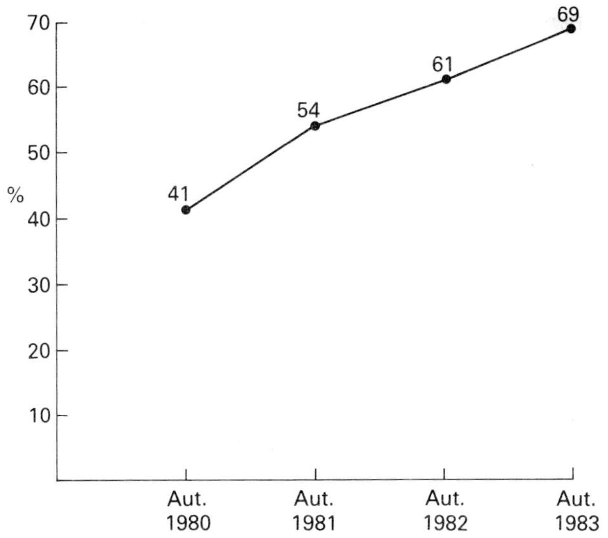

Figure 3.4 *Favourability towards British Telecom. Source: MORI.*

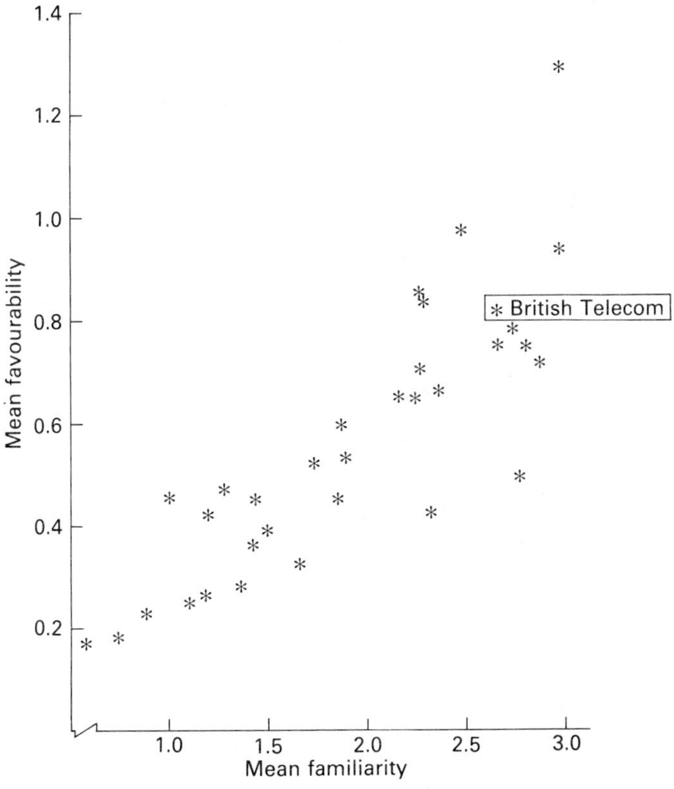

Figure 3.5 *Company familiarity and favourability, general public, Autumn 1983. Source: MORI.*

Marks and Spencer, Sainsbury and the Post Office. For example, awareness of British Telecom was high at 98 per cent and 75 per cent of the general public thought British Telecom provided a reliable service. Thirty-seven per cent thought BT had good future prospects, based on its technical sophistication, modernity and perceived profitability. Moreover, in the context of the privatisation debate in the media, over half the public questioned expressed an interest in buying British Telecom shares when the company was privatised. Such were the strengths of British Telecom a year before flotation. It was a strong and secure position for the company to be in and for the Government as the seller of British Telecom shares. The general image of British Telecom was secure and the corporate campaign could therefore build on British Telecom's strengths, while at the same time BT had a year in which to repair specific weaknesses. Public awareness of BT's business activities other than the telephone service—Prestel, radiopaging, telex, telemessages— was extremely low. On the financial side, British Telecom was perceived as poor at keeping costs down, and hence charging too much for its services; the view of management was negative and the company was perceived as rather too large and unwieldy. Such identified weaknesses as British Telecom had in its image profile would have to be specifically counteracted in the ensuing corporate campaign.

Defining the target audience

Thus, the MORI research established a detailed picture of how British Telecom was seen by the public, assessed its strengths and weaknesses, and provided an input into the strategy of the corporate and the separate offer campaign, which were complementary parts of the privatisation programme. However, a good general corporate image amongst the public was by itself insufficient to translate interest in British Telecom shares into purchase. Moreover, given the acknowledged lack of interest in share-ownership amongst the general public and an actual decline in private shareownership, additional research was required in order to discover, stimulate and motivate the target audience. A profile of the general public in terms of the disposition of their financial assets would help to give focus to this target and for this purpose National Opinion Poll (NOP) survey data was examined and formed the essential building block for calculation on the nature of the audience.

As can be seen from Table 3.1 a building society account was the most popular savings medium and over 14 million people had savings of over £1000 in their building society accounts. A close second was the public savings sector accounting for 17 million with savings in National Savings Certificates, Premium Bonds, Savings and Income Bonds, and money in the National Savings Bank, whilst a bank deposit account attracted some 14 million savings

Table 3.1 *Holders of financial assets 1983*

Category of holding	Holders as % population	Number million	ABC1	Profile 21–34	35–44	45–54	55+
Total adult population	100	41.0	39%	26%	15%	15%	33%
Shareholders	5	2.0	82	15	16	21	46
Unit trusts	3	1.2	74	17	20	26	36
Life insurance	55	22.5					
Bank deposit account	34	13.9	43	24	16	16	36
Building society							
Any high interest scheme	7	2.8	64	25	17	16	35
Building society							
£1000 plus accounts	35	14.3					
£2000 plus accounts	22	9.0					
£5000 plus accounts	9	3.6					
Total building society	52	21.3	49	30	16	15	30
Public sector savings	41	16.8	50	22	16	15	38
Index linked retirement certificates	3	1.2	60	1	4	7	87

Source: NOP Financial Research Survey, Great Britain

accounts. Almost at the bottom of the league table were stocks and shares. According to NOP, shareholders were in a minority. Only some 2.0 million people owned shares. The socio-economic group AB predominated, over 46 per cent of shareowners were over the age of 55 years and males predominated. From Table 3.2 it can be seen that ownership of stocks and shares is the preserve of higher income groups living mainly in the south-east of England.

It was evident that the private shareholder group would be a core target audience but that it was too small to meet the Government's wider share-ownership objective. Financial intermediaries as buyers on their own account were identified as another key target audience. Those with high disposable incomes would more than likely have matching savings and these could be a fruitful audience. Thus, given the high awareness of British Telecom and its strong public image, the adult population of the UK with savings of £2000 or more were set as the target audience for the British Telecom offer campaign. Research therefore identified the target audience for the British Telecom flotation in the form of concentric circles (Figure 3.6).

The British Telecom decision model

The next problem was to discover what criteria the target groups use in order to come to a savings or investment decision? A survey of the general public on the main criteria used in judging a potential investment was submitted to the

Table 3.2 *Profile of holders of stocks and shares*

	%
Percentage of the population	5
Sex	
Male	57
Female	43
Social class	
AB	56
C1	26
C2	11
DE	6
Age	
16–20	2
21–24	3
25–34	12
35–44	16
45–54	21
55–64	19
65+	27
Television area	
Scotland	11
Tyne Tees	3
Granada	11
Yorkshire	6
Central	10
Harlech	7
Anglia	7
London	26
TVS	14
TSW	5
Household income (of houses) up to	
£4499	6
£4500–£7499	9
£7500–£10 499	11
£10 500+	36

Source: NOP Survey 1983–84

marketing committee in March 1984. Fourteen factors were isolated as having a bearing on the general share investment decision. However, three major criteria were identified as the most important in relation to judging whether to buy shares specifically in British Telecom. Firstly, and perhaps not surprising, was British Telecom's profitability and therefore it followed that it would be a good investment. Secondly, was the perception of British Telecom's involvement in advanced technology and as an expanding company in a growth

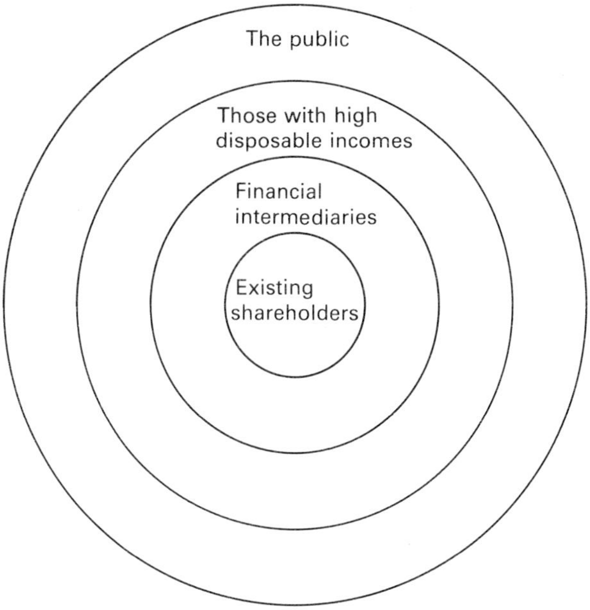

Figure 3.6 *The target audience for the retail marketing campaign.*

sector. Thirdly, was the perception that British Telecom might be even more successful when privatised and free from Government interference.

Research had highlighted and defined the target audience and the high standing of British Telecom's corporate image. It had focused attention on the high level of interest in buying BT shares and the rationale for that interest. However, at the same time, research had thrown up negative attitudes to ownership of shares *per se*. Shares were seen as a gamble, far removed from prudent saving. These findings determined the marketing approach. The retail marketing drive was to hinge on selling British Telecom, based on high awareness of its strengths and future rather than on selling shares *per se*.

But the real puzzle for the marketing committeee was how to translate interest in British Telecom to actual purchase of shares when the time came? Was there a way of predicting the final outcome? The answer came from Bob Worcester, long-established political forecaster and head of MORI. Based on his political decision and forecasting model but constructed differently, a conceptual model was devised (Figure 3.7) based on qualitative and quantitative research in Spring 1984 the key asset of which was to identify what differentiated the person interested in buying British Telecom shares from someone who was not interested. It was based on surveys undertaken to identify the level of interest—who was interested, why they were interested and how they would come to a decision on whether or not to invest in British Telecom. The research surveys also indicated what those interested in buying

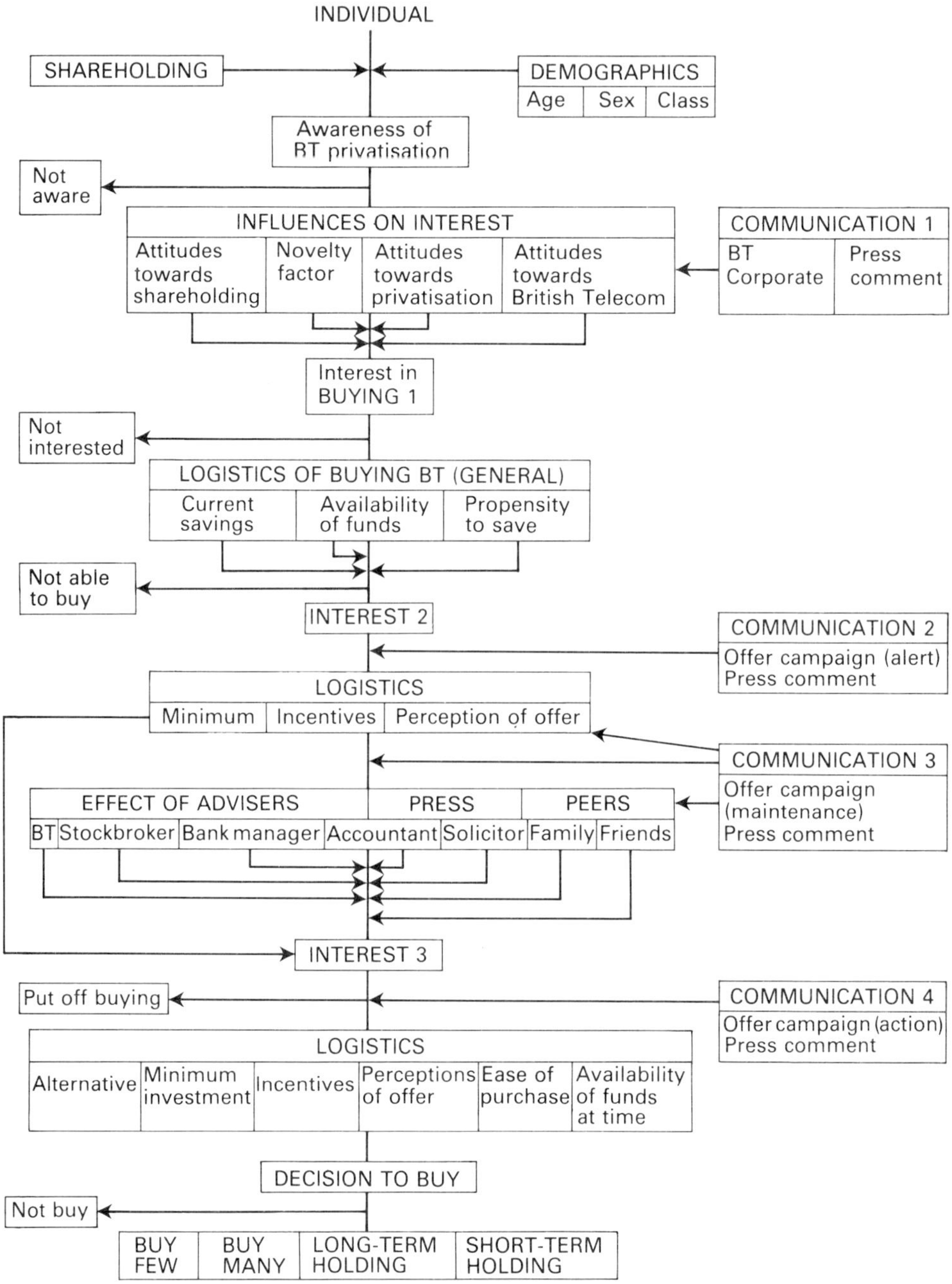

Figure 3.7 *British Telecom decision model.*

British Telecom shares would want to know, from whom, how much they were thinking of investing and the length of the investment commitment. Specific research studies were required to create the model and establish a baseline of figures for each box in the model. Subsequently, seven tracking studies then measured the change in attitudes and the impact of the offer communications. Suffice it to say that the model provided the blue-print for the communication strategy and its execution.

The British Telecom decision model was ready at the end of April, underpinned partly by the qualitative survey of that month. Public awareness of BT's forthcoming privatisation was high at 68 per cent and BT's strong positive image, in many respects higher than Sainsbury's and often equal with ICI's, was to be the foundation of the marketing appeal. Interest in BT shares was spread broadly across all sectors of the population and already 38 per cent of the public had said they were 'very' or 'fairly' interested in buying BT shares, partly on grounds of its past performance, its use of high technology, or its future prospects. Of lesser importance was support for the Government's policy of privatisation. In addition, though difficult to grasp or quantify, was the undoubted element of novelty and excitement about the prospective opportunity. In most circumstances investment is perceived as boring, but British Telecom was seen as big, safe and successful and the thought that one could make money undoubtedly existed. The majority of the sample perceived it as a long-term investment, the money for which would come from existing savings in bank and building society accounts. However, the amounts to be invested at this point in the campaign were rather low, with 74 per cent thinking of less than £500. As for sources of information, the public would use bank and building society managers, the newspapers, friends and family and British Telecom. The policy implication of the April quantitative study was striking: to be attractive on a wide scale, the terms of the BT share offer had to be perceived to give a better rate of return than a building society. This definition of expected income was more important than 'dividend payments', which for many of the public were incomprehensible. The April survey confirmed the favourable backcloth to the marketing campaign, deepened the research base and made a significant contribution to the construction of the British Telecom decision model.

In July, the first of the seven tracking studies was held in order to monitor changes in attitude towards the privatisation of British Telecom, to measure the impact of the communications programme, guide its future execution, and ultimately to predict the likely outcome of the share sale. Awareness of British Telecom was found to have reached 99 per cent during BT's corporate campaign under the slogan 'The Power Behind the Button'; awareness of BT's prospective privatisation averaged 71 per cent with scores of 89 per cent among a key target segment: AB men. By age groups, awareness at 80 per cent was highest among the 45–54 age group. Combining those who said they were interested and fairly interested it was possible to assess 'interest' in actually purchasing BT shares at 18 per cent, spread fairly evenly among socio-economic groups AB, C1, C2 and DE. Reasons for buying British Telecom shares boiled down to the perception that it was a successful company of the future and a secure investment. Support of privatisation policy just marginally exceeded attraction by the novelty of the concept. At a minimum investment level, 48 per cent of those 'interested' in buying BT shares said they were certain or likely to buy, whilst support for the telephone voucher and bonus issue was unevenly spread at 36 and 54 per cent respectively. Instalment payments for many of the sample would not affect their decision to buy British Telecom shares.

These were some of the main findings in the first tracking study conducted in July. However, it must be appreciated that the surveys were in fact far more detailed. For every measurement (for example, of awareness of privatisation) several other measurements were monitored. Awareness was broken down by socio-demographic-economic factors, such as class, sex, age-group, readership profiles, trade union membership, ownership of a telephone in the home, location by television region and current financial holdings. The same breakdown was applied for other major questions, and the separate comparisons of perceptions regarding an investment in British Telecom and a building society were explored.

The second tracking study, conducted at the end of August, was timed specifically to measure the impact of the first eight days of the 'Alert' phase offer advertising, which began on 20 August. As can be seen from Figure 3.8 general interest in buying shares has risen by 6 per cent to 24 per cent, although the proportion 'very interested' remained constant at 5 per cent. Certainty of buying remained at roughly the same level as it was in late July. Six per cent of those who were interested said they were certain to buy the shares and this represents 1.5 per cent of the population or 630 000 people. The average amount that those who were certain to buy said they would invest was over £450. The research also found a significant change in the composition of interest by social class. Whereas in July, interest in buying BT shares had been evenly spread across the social classes, in August, there was a clear middle-class and working-class division. Whilst 30 per cent of socio-economic groups AB and C1 were interested, only 23 per cent of C2s and 17 per cent of DEs were found to be so. There were few changes during August in the reasons for interest or lack of interest in the shares. Some 36 per cent of the general public said they had seen the offer advertising in the previous eight days; the figure was 50 per cent for those interested in the shares, perhaps because the advertising was more relevant to them. Furthermore, the advertising had been well received by those who had seen it, with nearly a third saying it was educational, and just around a quarter saying they found it interesting and informative (Figure 3.9). There were also sharp rises in August in the proportions associating BT with being a good company to own shares in (up from 27 to 35 per cent) and being

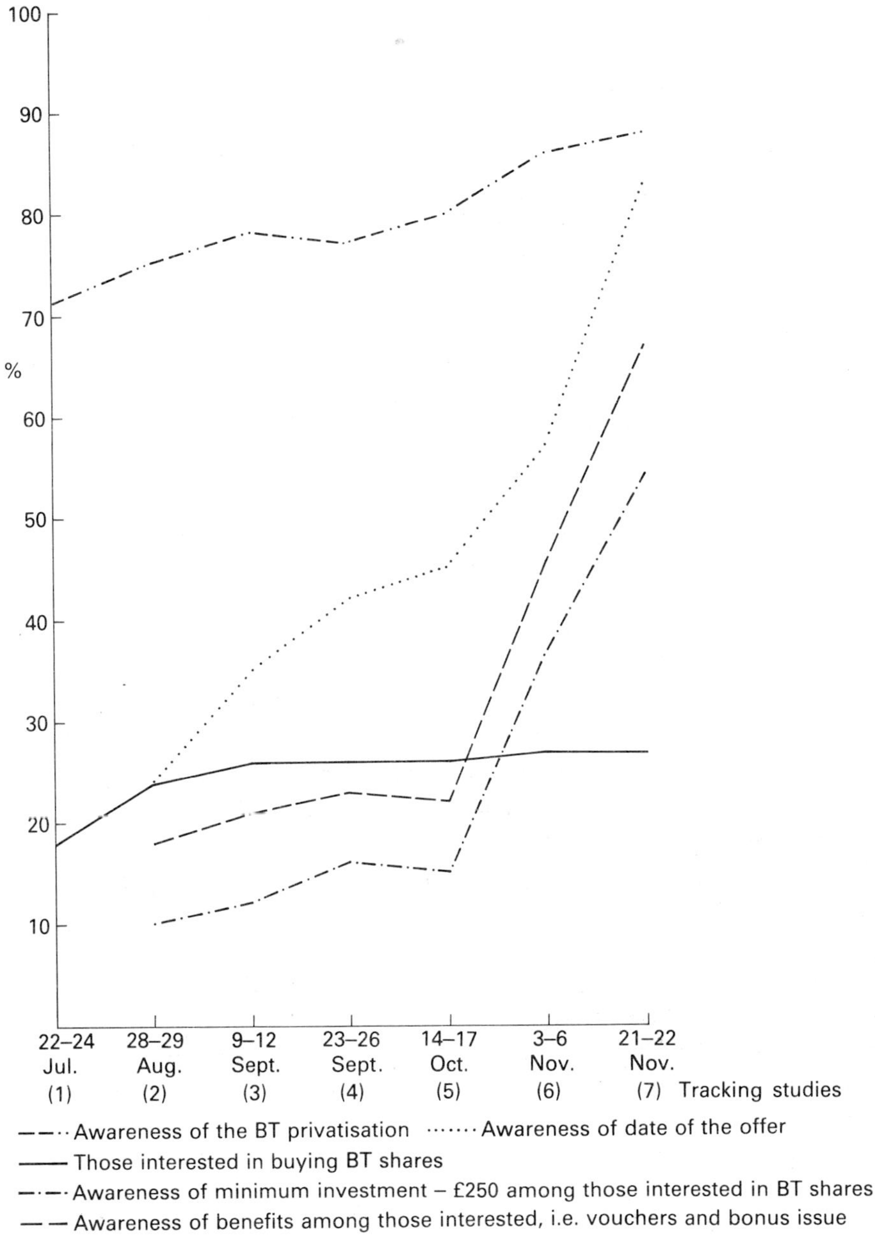

Figure 3.8 *Tracking studies 1–7 measuring the impact of the marketing drive. Source: MORI.*

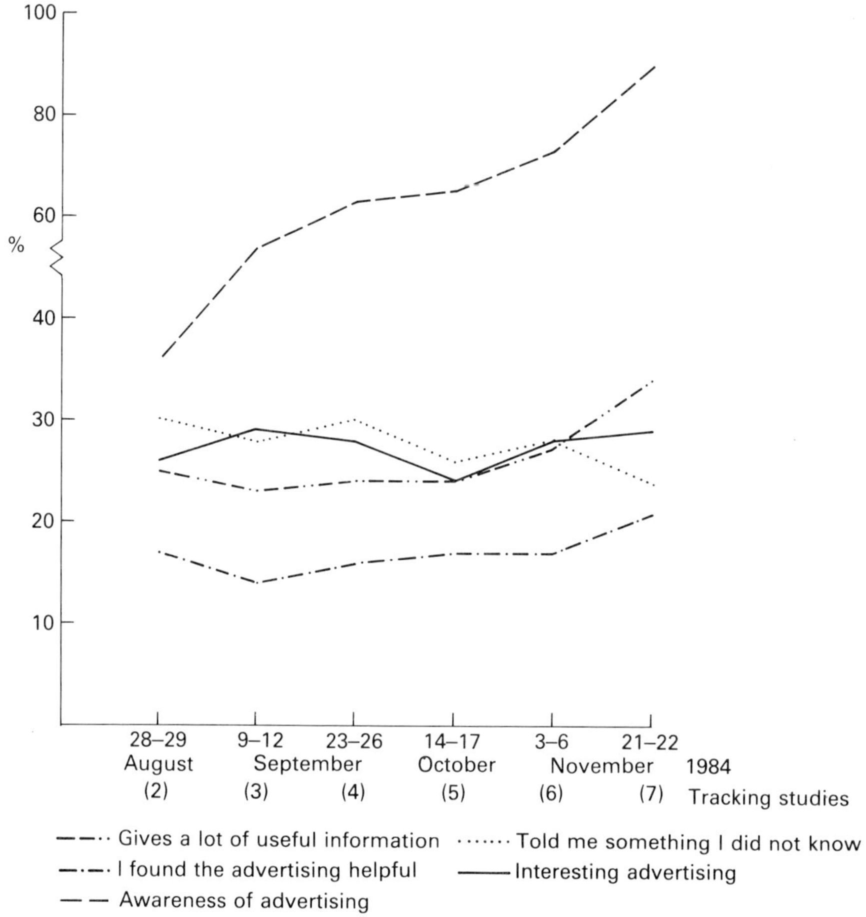

Figure 3.9 *The public's views of the offer advertising. Source: MORI.*

a safe investment (up from 28 per cent to 35 per cent). However, although awareness of the actual privatisation was very high at 75 per cent, knowledge about the timing of the sale, the benefits for individual shareholders and the minimum level of investment was low. These information gaps revealed by the research, led to press advertising giving specific information on these items. Finally, the survey revealed that the public were still undecided on the relative investment merits of British Telecom shares and a building society account.

The third tracking study conducted in the second week of September measured changes in attitude during the main part of the 'Alert' phase advertising. Interest in buying BT shares remained fairly constant at around 26 per cent of the population. Seven per cent said they were 'very' interested in the shares. Certainty of buying, like interest, remained at roughly the same level—1.8 per cent of the population or 750 000 people. Also rising was the

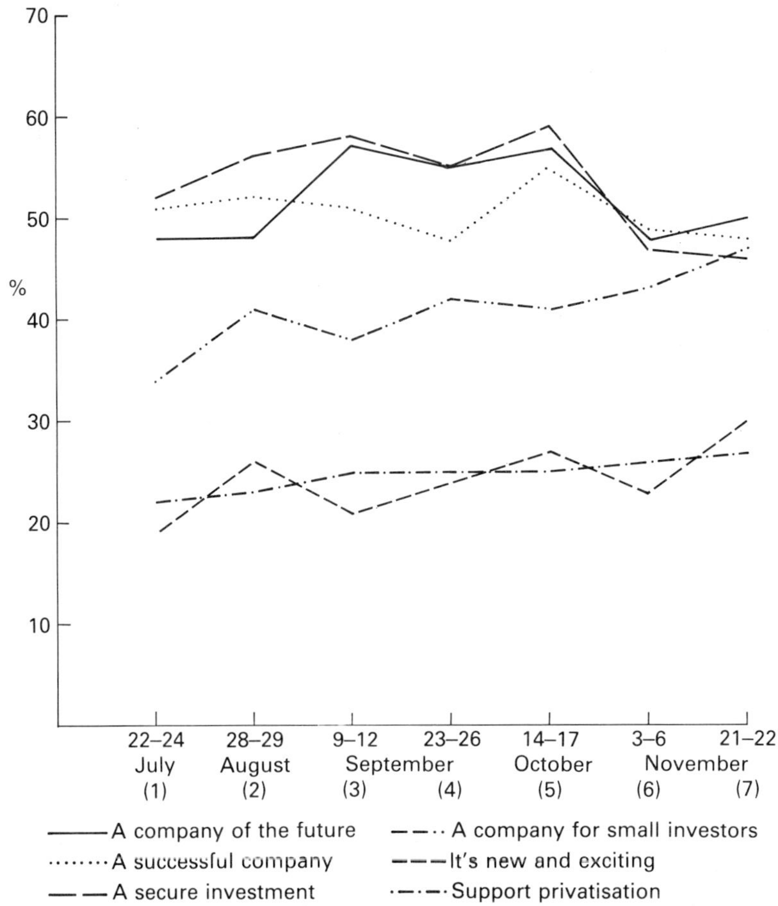

Figure 3.10 *Tracking studies 1–7 among the general public: reasons for interest in BT shares by those very or fairly interested. Source: MORI.*

average amount that those certain to buy said they would invest, now up by around £100 to £550. There were no real changes during the first half of September in the reasons for interest or lack of interest in BT shares. However, among those interested, those saying BT is a company of the future rose by 9 points from 48 to 57 per cent; BT's image as a good and safe investment also rose (Figure 3.10). Awareness of the privatisation (78 per cent) and of the advertising (54 per cent) was very high. However, knowledge of the details of the offer continued to remain at a low level (Figure 3.8).

The fourth tracking study, conducted in late September, measured public response at the beginning of the 'maintenance' phase offer campaign. Whilst interest in BT shares remained constant at around 26 per cent of the population, the proportion 'very interested' rose by 1 per cent to 8 per cent and certainty of buying rose to 2.6 per cent of the population, equivalent to over one million people. Similarly, the average amount to be invested rose an additional £100 to over £674. During September BT's overall image declined slightly, possibly as a result of the ending of the corporate campaign and the onset of the decay factor. The proportion associating BT with being profitable, efficient and a high technology company fell. Finally, knowledge of the other details of the offer with the exception of its timing remained very low.

The fifth tracking study covered the first two weeks in October, a time of increasing newspaper editorial on British Telecom's share offer in the financial pages and concurrent with the 'Maintenance' phase of advertising. Most of the measurements remained constant. For example, interest in buying BT shares stood at 25 per cent of the population. Eight per cent said they were very interested in the shares. Certainty of buying remained roughly constant at 2.2 per cent of the population or some 900 000 people. The level of interest was still higher among the middle classes. There was likewise no change in the overall pattern of reasons for interest in BT shares. The major reasons for interest remained the success, security and future prospects of British Telecom, whilst lack of money remained the major factor for lack of interest in share purchase—an understandable reason, for at least 20 per cent of the population had no savings at all. Finally, there was a continued decline in BT's overall image, first noted in September, and with it a reduction of the average sum to be invested to around £630.

The sixth tracking study, the last before the pricing decision, was conducted in early November, and measured the public's views during the latter part of October, which was accompanied by the final stage of the 'maintenance' advertising and was a time of increasing coverage of the share offer in the press and television. Many of the measurements remained stable. For example, interest in buying BT shares stood at 27 per cent, around the same level as in late August. Certainty of buying had, however, increased slightly and represented some 1.2 million people but the average amount to be invested had declined to just over £500. Likewise the major reasons for buying BT shares—that it is a successful, secure company of the future—showed a decline during October. However, BT's overall image remained roughly constant in late October and the decay factor seems to have been arrested. By contrast, awareness of the flotation advertising was up from 65 per cent in mid October to 73 per cent in early November, and awareness of the privatisation rose to 86 per cent. As to the impact of the offer advertising campaign, an impression was finally being made. For example, awareness of the various benefits was rising.

The seventh and final tracking study among the general public was conducted on 21 and 22 November, covering the period immediately after the publication of the prospectus. It provided the information, based on a sample survey, that 1.7 million had applied and a further 1.5 million were certain to buy, a total of 3.2 million people. In addition the average amount that those who said they were certain to buy or who had already invested nearly doubled to some £900. In the event, 2.3 million applications were received, of which nearly 1.5 million opted for an investment of less than £800. Such a result was far greater than the earlier expectations of the marketing committee but lower than the indications given by the mid-offer research. Perhaps surveys of this kind can only measure broad current attitudes and indicate variations in these attitudes from earlier findings, giving a broad indication of the likely trend. Nevertheless, the MORI research findings were invaluable to the marketing strategists, not only as a guide to the content of the advertising campaign but also as one of the few in-depth monitors of the effectiveness of advertising campaigns. From Figure 3.8 two striking points emerge. One is the length of campaign which elapsed before some really significant movements in knowledge occurred. For it was only during the second half of October that awareness of the minimum level of investment, the details of the voucher scheme and bonus issue and the timing of the offer began to take-off. More gradual was the steady incremental gain in the measurements for awareness of the flotation. However, from the last week of August the proportion 'interested' in buying BT shares remained more or less constant at some one in four of the population and moreover the increase in certainty of buying BT shares took place within this group. This of course carries the implication that for much of the campaign it was preaching to the already converted. From Figure 3.9 it can be seen that views on the advertisement campaign remained almost constant with little movement, apart from the gradually rising trend of awareness of advertising.

ESTABLISHMENT OF A RETAIL DISTRIBUTION NETWORK

According to Inland Revenue figures 1.6 million individuals or 3.5 per cent of the adult population owned shares before the BT offer. This was not always the case, for the proportion was once much higher but the pattern of security ownership in the UK has changed dramatically over the past two decades. Dealings on the Stock Exchange are dominated by institutional investors, both on a day-to-day basis and in the long term. Financial institutions such as pension funds and insurance companies provide an overwhelming proportion of new capital raised on the Stock Exchange. Thus there has been a dramatic shift in ownership of equities, away from the private shareholder to the financial institutions as can be seen in Figure 3.11. Between 1963 and 1981 institutional holdings of UK ordinary shares rose from 28 per cent to 59 per cent, while individual holdings fell from 54 to 28 per cent. Furthermore, since Mrs Thatcher's Government took office in 1979, private individuals have been net sellers of more than £13 billion worth of shares, mainly to insurance companies and pension funds. The figures demonstrate the high concentration of share-ownership, serviced in the main by 201 UK Stock Exchange broker firms and local bank branches. Such traditional investors would no doubt be a core target group for British Telecom shares but their numbers were diminishing. This is not the place to rehearse the well known arguments for the demise

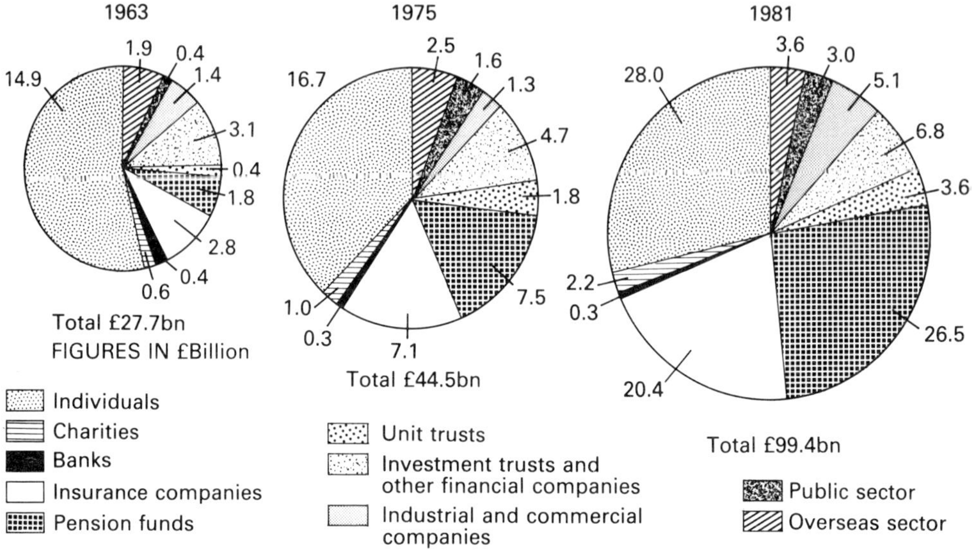

Figure 3.11 *Distribution of shareholding between categories of beneficial shareholder by market value. Source:* Financial Times, *15 November 1983.*

of the small investor, but they do include high dealing costs and stamp duty. The more interesting and fundamental question is why have not a substantial proportion of the holders of 38 million building society accounts and over 12 million home owner-occupiers acquired the habit of investing in equities? Clearly there are many potential investors for at the end of 1984 building societies alone held over half or £90 365 million of personal sector liquid assets and during the year building societies attracted nearly two thirds of new savings. A partial explanation lies in the distribution network of stockbrokers.

It is perhaps true that the existing stockbroker network is an effective mechanism for buying and selling shares in the secondary market as well as for the buying of new shares, but the scale of the Telecom new issue exposed the weakness in the distribution pattern. The size of the issue was so large that neither the traditional market for shares nor the stockbroker network were adequate. Enquiries exposed an immense gulf between the customary pattern and what the Telecom issue required—that is, a national distribution and advisory service. No UK stockbroker has a retail network like his American counterpart with an established 'high street' presence. Similarly no retailer in the UK has followed Sears Roebuck, the biggest retailer in America in selling shares in its stores, but imagine the impact if Marks and Spencer sold shares!

Moreover, the accessibility of stockbrokers is limited in other ways. Whilst small stockbroking firms based in the provinces and London offer services to private clients, the majority have concluded that the commission income from private client transactions cannot justify the administrative cost. Instead they have concentrated on a superior and more lucrative market segment, the institutional investor. In these cases stockbrokers have done nothing to deepen or widen their market. If anything, they have repelled it, requiring a new potential client to supply references, a demand not made by building societies, pension funds, insurance companies or clearing banks. Finally, Stock Exchange members will admit to the prevailing myth that the Stock Exchange is a rich man's club with open doors only to the wealthy for it has in fact a very real foundation. In Spring 1984, according to *The Times*, minimum investment levels for private clients were high, at around £50 000 for Phillips and Drew and £40 000 for Hoare Govett for their advisory services and £6000 for their unit trust portfolio management services. Added to this reality, there are beliefs and images which may have a very strong impact on behaviour. Research commissioned in 1984 by stockbrokers Montague Loebl Stanley showed that even most of those people already using a broker believed that at least £20 000 was needed before a London broker would accept a client. Of lesser importance is the fact that few stockbrokers advertise for private clients; many have an unapproachable image and a significant proportion have deliberately eschewed the smaller investor. As the results of the Telecom issue reveal, 66 per cent of all small investors buy their shares through the local bank branch and one million successful applicants applied for between 200 and 400 shares, representing a total investment of £260 and £520 respectively or between one sixth and one third of the figure which many brokers require for a single deal.

Given the gulf between British Telecom's large-scale marketing objectives and the restricted range of stockbroker clientele it was necessary to use all the resources of advertising, public relations and financial intermediaries to reach new markets and provide fully-briefed sources of information and advice. This required the involvement of a huge and unique intermediary network.

Professional financial intermediaries

To achieve its desire for a wider spread of share-ownership the Government had authorised the largest marketing campaign ever for a new issue to inform and create public interest in the share sale. Market research undertaken by MORI in April 1984 demonstrated the public's need for information and advice on British Telecom and its share offer. Only 2 per cent of the public expected to contact a stockbroker for information and advice, thus confirming the structural, organisational and motivational weaknesses of the existing stockbroker mechanism beyond its traditional market-place. Amongst the general public those seen as the most important sources of information were bank managers, building society managers and the press. It is interesting to note that both non-share-owners and share-owners attached equal importance to the bank manager as a source of information. However, a sizeable proportion of the public indicated that they would turn to independent professional financial intermediaries such as insurance brokers, accountants and solicitors. Moreover, research indicated that 94 per cent of those interested in buying British Telecom shares would seek advice on whether to buy the shares. Of those 'fairly' or 'very interested', 62 per cent indicated that they would seek advice from their bank managers and 33 per cent said they would seek advice from accountants, solicitors and insurance brokers. Furthermore, when questioned on how they would like to buy their shares the answers indicated a mix of distribution channels ranging from over-the-counter at the post office (the most popular method) to bank branches, British Telecom offices and showrooms, building societies, stockbrokers or by filling in an advertisement coupon. The answers suggested the basis of the distribution strategy. Banks, post offices and the press were to be key information points and the disseminators of the application form. Stockbrokers were to be a source of information and advice in their traditional capacity and customary market. Financial intermediaries such as accountants, solicitors, insurance brokers and bank managers would need to be involved partly out of public demand and partly in order to penetrate the retail market, based on their national representation and numbers. In 1984 it was estimated that there were over 33000 such financial intermediaries—14000 accountants, nearly 11500 solicitors and at least 7000 insurance brokers. This considerable force could be harnessed as a source of information, advice and as receiver of applications from the public.

However, there were unexpected difficulties. MORI research findings indicated some disturbing features. First of all, over 70 per cent of bank managers, 80 per cent of accountants and 95 per cent of solicitors assigned little importance to an advisory function and the majority of them expected little general public demand for advice. Furthermore, apart from stockbrokers, the vast majority of intermediaries would wait for clients to approach them rather than take an active stance. This passivity of financial intermediaries in itself was rather surprising and was reinforced by another MORI finding that a significant number of intermediaries did not wish to receive many requests for advice on British Telecom shares from people with less than £1000 to invest and most expected an investment of over £500. Given this context of passivity in relation to the market-place, of relatively high investment thresholds and of the gap between consumer expectations from financial intermediaries and intermediaries' own perceptions about their role it became clear that a separate marketing campaign directed at financial intermediaries was necessary in order to gain the interest and support of this sizeable national group. An essential strand was the provision of information through a variety of avenues including a central enquiry office, a direct-mail campaign, publicity material, advertising in trade magazines, a series of briefing seminars on the British

Telecom train and roadshows to reinforce their exposure to the general marketing campaign in the press, television, radio and posters. It was hoped that such measures would at least raise awareness, and perhaps encourage involvement in the campaign, either by persuading them to tap the traditional shareholder base or to secure new money from the first-time investor. It might simultaneously encourage financial intermediaries to invest on their own account. Furthermore, to encourage financial intermediaries to deal with strangers for relatively small amounts there was the additional incentive of commission payments. Solicitors, accountants, licensed deposit-takers and dealers, insurance brokers, banks and stockbrokers were entitled to various levels of commission on successful applications processed by them. A stockbroker submitting successful applications up to £10 000, bearing a single broker stamp, was entitled to receive the full 2 per cent commission, and an additional half per cent on the excess over £10 000. Banks submitting applications direct were to receive 1.5 per cent on the first £10 000 or, if submitting through a stockbroker (whereby the form carried two stamps, one from the broker, one from the bank) commission of 1.625 per cent was payable. For other financial intermediaries all applications would have to be submitted through a stockbroker; the latter would be paid the whole commission but be obliged to allocate to the intermediary 1.25 per cent on successful applications up to the value of £10 000 and 0.25 per cent for sums above that figure.

The regional stockbroker co-ordinator

It became clear that in order to mobilise and control this army of intermediaries it would be necessary to group them by regions with a central control centre, in the form of a stockbroker who with his local connections would act as a collection and distribution point for information within his region. In March 1984 22 regional co-ordinators had been appointed to cover the UK (Table 3.3). Because of the dominance of London the counties nearer to the capital did not contain broking firms of the appropriate size and status to fulfil the regional co-ordinator role. Even areas with a strong regional identity, such as East Anglia, lacked both the proper nucleus essential to a region and suitable broking firms to take on the task of the regional co-ordinator role. It was decided therefore to allocate responsibility for all the Home Counties to a large London broker, Quilter Goodison, with a strong private client list. The counties to the west of London, Berkshire, Oxfordshire and Wiltshire were grouped in a rather amorphous region centred on Reading, whilst London was the responsibility of the four members of the 'charmed circle', Hoare Govett Ltd., Cazenove & Co., Scrimgeour, Kemp-Gee & Co., and De Zoete & Bevan, lead brokers to the British Telecom share issue. In some cases where the broking community was stronger, such as Edinburgh, Glasgow and

Table 3.3 *Regional co-ordinator stockbroker firms*

1. London
   — Hoare Govett Ltd
   — Cazenove & Co.
   — Scrimgeour, Kemp-Gee & Co.
   — De Zoete & Bevan
2. Sussex, Kent, Surrey, Hampshire, Dorset and Isle of Wight, Essex, Suffolk, Norfolk, Cambridgeshire, Hertfordshire, Bedfordshire, Buckinghamshire and Northamptonshire
   — Quilter Goodison & Co.
3. Devon and Cornwall
   — Westlake & Co.
4. Avon, Somerset and Gloucestershire
   — Stock Beech & Co.
5. Berkshire, Wiltshire and Oxfordshire
   — Heseltine Moss & Co.
6. Dyfed, Glamorgan, Gwent and Powys
   — Lyddon & Co.
7. Shropshire, Staffordshire, W. Midlands, Warwickshire, Hereford and Worcester
   — Albert E. Sharp & Co.
   — Smith Keen Cutler
8. Nottinghamshire, Lincolnshire and Leicestershire
   — Hill Osborne & Co.
9. Clwyd, Cheshire, Gwynedd, Merseyside and Isle of Man
   — Tilney & Co.
10. Greater Manchester, Derbyshire and Lancashire
    — Henry Cooke Lumsden & Co.
11. Yorkshire and Humberside
    — Wise Speke & Co.
    — Stancliffe Todd & Hodgson
12. Northumberland, Cumbria, Tyne and Wear, Cleveland and Durham
    — Wise Speke & Co.
    — Stancliffe Todd & Hodgson
13. Borders, Lothian, Fife, Central Scotland and Edinburgh
    — Bell, Lawrie, Macgregor & Co.
    — Wood, Mackenzie & Co.
14. Dumfries and Galloway, Strathclyde and Glasgow
    — Speirs & Jeffrey
    — Penney Easton & Co.
15. Grampians, Highlands and Tayside
    — Parsons & Co.
16. Northern Ireland
    — Wm F. Coates & Co.

Birmingham, it was decided to appoint two firms to share the role and in these there was no lead firm nominated. In the case of the North East and Yorkshire the same two firms were appointed for each region and worked as a pair in each, both having offices spread throughout the regions. Northern Ireland was covered by a Belfast broking firm, a Bristol firm covered Avon, Somerset and Gloucestershire whilst a Plymouth firm was appointed to cover Devon and Cornwall.

To establish more regions than sixteen would have maximised local loyalty but produced a management problem at the centre, whilst fewer regions would have meant a loss of local identity. The number of regions was settled as a compromise and when the lead brokers to the issue sat down to discuss regional brokers for them, it became clear that these were almost self-selecting in the sense that there was one obvious firm, the leader in its region.

Regional co-ordinators were required to contact all financial intermediaries in their area and build up a mailing list not only for the distribution of information and publicity materials, prospectuses and application forms but to propose guests for the BT train and roadshow briefings and define the audience for the regional co-ordinator's own marketing drive in his area. Basically, it was the job of the regional co-ordinators to 'ginger up' financial intermediaries in their area and to act as a focal point for their region, and to report to the lead brokers, Hoare Govett.

The intermediary direct-mail programme

The British Telecom Share Information Office and Mail Marketing Ltd. fulfilled a dual purpose. Not only was it the nucleus for information on British Telecom and the flotation for the retail sector, it was also the nerve-centre for the direct mail operations targeted at financial intermediaries.

Research had highlighted the importance of financial intermediaries as a source of advice. Therefore it was imperative that the 201 stockbroker firms, 14200 accountant firms, 11430 solicitor and 7000 insurance broker firms be kept fully informed about the company and the offer details to enable them both to inform existing clients and to advise new ones. Education, information and active support for the new issue were the aims of the intermediary marketing campaign.

One route to these objectives was the direct-mail campaign. Between late August and 19 November intermediaries were sent four mailings. The first from 20 August coincided with the appearance on national television, press, posters and radio of the flotation advertising with the theme 'You can share in British Telecom's future'. It was designed to secure intermediaries' interest and provide a basic briefing. All 35000 intermediaries and 201 stockbroker firms received a package with covering letter from John Harben with the special box number and telephone number reserved by Telecom Tan and Mail Marketing for intermediaries' queries. The letter anticipated massive public interest in the issue and the need for intermediaries to be in a position to advise the public. Thus intermediaries received a copy of the share information fact sheet in a question-and-answer format reminding them of their likely role, tempting them with details of the scale of commissions on successful applications, explaining the operation of the Government-financed incentive schemes and above all reassuring them of public interest. It did this by drawing their attention to the planned mass distribution of the prospectus via bank brances, post offices, advertisements and the wide-scale distribution in leaflet form of the unique abridged prospectus. Intermediaries were informed that a Pathfinder prospectus would be sent to them three weeks before the issue and that around £250 would be the minimum total investment. For further information, additional copies and a video on BT's activities intermediaries could contact the British Telecom Share Information Office (at Bristol) or contact their regional co-ordinator, The first mailing included a copy of Sir George Jefferson's letter to staff of May 1984 outlining the special provisions for employees to buy shares. In addition intermediaries received the information pack, which was available to all enquirers at the British Telecom Share Information Office (BTSIO) and which included the Stock Exchange booklet (a guide to buying and selling shares); the A4 information fact sheet, and the British Telecom 'Power Behind the Button' brochure. By 9 October the BTSIO had received over 1100 requests from intermediaries for more copies of the literature and merchandise.

The second mailing to intermediaries in October was again accompanied by a cover letter which reassured intermediaries of widespread public interest since public enquiries to the BTSIO had reached many thousands a day. More information on BT was provided in an 'A to Z' guide describing the company's products and services, and to sustain intermediary interest they received an advertising-campaign 'presenter'—a booklet containing samples of current press, television and poster advertising as well as a sample media schedule designed to show the forthcoming heavy-weight advertising campaign. By 6 November 4696 intermediaries had contacted the BTSIO for further copies of the literature.

On publication of the Pathfinder prospectus on 26 October intermediaries were to receive a third mailing, warning them that the offer was due the following month and giving them details of the timetable. Once more intermediaries were reassured of public interest in the flotation for MORI research showed that one quarter of the adult population were thinking about investing. A completed specimen application form gave guidance on the procedure for filling in the actual application and finally, intermediaries received, like the general public, a BT share offer information sheet concerning the flotation. By 13 November 5031 intermediaries had contacted the BTSIO.

The fourth and final mailing began on Monday 19 November just after the announcement of the BT share price. This was the culmination of a concerted marketing drive directed at the intermediary network and the retail sector and was principally concerned with briefing intermediaries on the logistics of prospectuses and applications. Each letter was accompanied by five copies of the abridged or 'mini' prospectus. The completed application forms of their clients could be hand-delivered to one of 21 receiving banks or to one of six postal return addresses according to the alphabetical system matching each applicant's name with one or other of the banks (Table 3.4).*

* Dealings in BT shares would commence on 3 December 1984.

Table 3.4

First letter of surname	Receiving bank
A to CL	Bank of Scotland
CM to G	Barclays Bank
H to L	Lloyds Bank
M to PH	National Westminster
PI to S	The Royal Bank of Scotland
T to Z	Midland Bank

The offer would close at 10 a.m. on Wednesday 28 November and in order to meet that deadline intermediaries were warned to use first-class mail post at the latest on 26 November. At the same time intermediaries were warned of the Government's decision that no commission would be paid on suspected multiple applications.

The British Telecom exhibition train and PLC coaches

The idea of direct face-to-face contact with provincial financial intermediaries emerged from the marketing committee. One of the vehicles (in a literal sense) was to be the BT exhibition train. For the last five summers BT had customarily employed a British Rail train as a mobile exhibition to show off its equipment to existing and potential customers. The marketing committee added two coaches to provide a mobile conference hall and hospitality area and added two stops to the usual itinerary (Table 3.5). The two additional coaches, named

Table 3.5 *British Telecom PLC Train Itinerary, 1984*

Date	Stop
29 May	Truro
30	Rehearsal in Truro
31	Plymouth
1 and 4 June	Bristol
5 and 6	Birmingham
11	Liverpool
12 and 13	Manchester
18	Glasgow
19 and 20	Edinburgh
21	Aberdeen
25	Newcastle
26 and 27	Leeds
29	Nottingham
3 July	Norwich
5	Milton Keynes
6	Brighton
9	Southampton
10	Cardiff
11	Oxford

the 'PLC' coaches, would present British Telecom in the context of privatisation to local media and intermediaries. The objectives were clear: to meet and brief key audiences at each stop, and to gauge particular communications needs and opportunities, highlighted by questions from the audience and comments over coffee, lunch and tea. It was hoped to fire the audience with enthusiasm about BT as a potential investment by displaying its range of activities and products. The train also tested the 'market power' of the 22 stockbroking firms appointed by Hoare Govett as regional co-ordinators. The regional co-ordinators were responsible for choosing the financial intermediaries to be invited. Guest lists were drawn up and submitted to Hoare Govett, and invitation cards featuring the ubiquitous telephone whose receiver was transformed into a train were sent out by the regional co-ordinators in the second week of May. The train placed regional co-ordinators as the generators of later marketing drives and as the focus for financial intermediaries in their respective areas. Equally the train demonstrated the serious intention of the new-issue organisers to run and co-ordinate a retail marketing campaign and at the same time reassured the financial institutions in the City of London that this was to be a national campaign with all the benefits of national participation.

Thus, on a clear, crisp morning on 30 May 1984 a seven-coach train pulled out of a siding in Truro on a 3500 mile, six-week journey to begin the education of financial intermediaries and stockbrokers on the company, British Telecom, and its flotation. It was the first outward sign of the flotation marketing programme, and timed to coincide with the 'account mailer' to be sent to 20 million British Telecom customers over the next twelve weeks.

In each city, up to six presentations were made to groups of 30 intermediaries in the two PLC coaches of the train. A typical three-session day began with the morning seminar at 9.30 and ended with coffee at 11.15 a.m. The mid-day session between noon and 2.30 p.m. included a standard lunch. The afternoon session including tea began at 3.55 and ended at 5.30 p.m.

On board the BT train were representatives of most of the key participants in the flotation exercise. British Telecom itself was represented by Frank Lawson (then director, special marketing projects), the relevant local BT manager, Antony Stanley-Smith (from Kleinwort Benson), Michael Francis (a partner of Hoare Govett), and representatives from S. G. Warburg. The 'hosts' were a distinguished and impressive team and reinforced the impression of a serious intention to mass-market the BT issue.

Guests were welcomed to the seminar coach and treated to an hour-long presentation. A brief welcome and introduction by the regional co-ordinator was followed by a 15-minute film 'The Power Behind the Button'. This British Telecom was not just a telephone company but a high-technology international company. To press the point home a British Telecom regional general manager drew attention to BT's local presence and activities. The presentation by Frank

Lawson confronted the image perception of BT directly. BT was not a public utility, it was a high-technology company whose business was to 'to provide customers with the means of assembling, exchanging and processing information based on the convergence of telecommunications, computers and broadcasting'. He stressed the switch in management thinking towards customers. New growth opportunities lay in the electronic office and office automation. The network was being improved with the introduction of System X and other electronic exchanges to give faster, more reliable and cost-efficient connections for customers. The general network was becoming more flexible and already nearly 70 companies provided over 200 value-added network services. For the business community high-speed data transmission was already here and for the roving businessman, portable communication facilities in the form of carphone, cellular radio and bleeper paging provided satisfaction of customer needs and opportunities. Whilst the future looked rosy, the present was more than satisfactory. Already BT had a turnover of nearly £7 billion, a growth rate of over 7 per cent in turnover a year; international traffic growth of between 11 and 20 per cent a year and over 29 million phones *in situ*. A marketing drive to sell new, stylish, sophisticated phones from high-profile BT shops and franchises in the High Street was being matched with a call stimulation campaign, for if each UK customer was to make only one extra trunk call a day, BT's revenue would increase by several hundred millions a year.

Mr. Lawson's speech was followed by Michael Francis on the subject of marketing the offer to the general public. The ultimate objective of the campaign was widespread share ownership—but in the process there were several targets to be aimed at. It was essential to make the flotation a national event. Moreover, unlike all previous new issues, this one was not to be shrouded in secrecy until impact day. The details of the offer were to be heavily publicised and advance notice given of minimum investment levels, shareholder incentives, together with the unique 'mini' prospectus. British Telecom was to be promoted as a serious investment opportunity for all, but for the financially sophisticated three stockbroker reports were to be made available. In addition the BT annual report and accounts for the year to 31 March 1984 would be available on request from 19 July. The application period, given the scale and nature of applicants and the logistics involved, would be extended from the customary 3 to 8 days. The general features of the prospectus, excluding the price, would be made available in good time in the form of the Pathfinder prospectus in order to give investors, advisers and financial journalists ample opportunity to take in the information, which would be considerable. Later, Mr. Francis was able to tell his audience that a series of provincial roadshows was planned for each region in October. Finally, to increase intermediaries' interest and engage their active support, the Government's commission structure was announced. Guests received for reference a comprehensive package of supporting information. Each session ended with formal questions-and-answers chaired by Frank Lawson. Questions on BT ranged from the general, such as the effect of the RPI−3 formula, the question of future Government control, the attitude of the BT unions to privatisation, the ability of BT in a competitive environment, and the effect of BT's heavy investment programme on its ability to pay dividends—to specific enquiries, such as loss-leaders and the questions of where BT would be found on the Stock Market page of the *Financial Times*! On the actual offer itself, the panel were asked for additional details on the rental-voucher scheme, the price and dividend yield, the split of shareholders between institutions and private individuals, the importance of overseas capital raising, underwriting procedure and the nature of the issue. Was it to be a fixed price or tender offer? Many of these questions were unanswerable at this stage. Following the formal question session, further opportunities for discussion arose during the hospitality session where guests mingled with their hosts.

At the end of each day, media press relations activity came into its own. Invited by Dewe Rogerson, writers and broadcasters from regional and local media came on board for informal briefings to give provincial media first-hand coverage instead of syndicated material from London, and to encourage local coverage of British Telecom. Journalists received the information packs and live interviews and photocall sessions were arranged. Once the media briefing was completed the train left its siding for the next destination, another city, but in the place it had just left the day invariably began with substantial media coverage.

In total 1300 financial intermediaries attended a session on the PLC train and 50 per cent applied to the BTSIO for further literature. MORI research findings suggest that the train had clearly increased intermediaries' awareness and understanding both of BT and the general features of the share offer. Guests freely admitted that their view of BT had altered and improved and many said they would spread the word to clients and colleagues. And those intermediaries who had attended the train session were slightly more likely than their 'non-train' colleagues to see themselves as important in advising the general public on whether or not to buy shares in BT.

The train therefore succeeded in informing intermediaries and motivating them to perform the role allocated to them in the offer. For the regional co-ordinators it established a liaison and dialogue with financial intermediaries in their area, and incidentally stimulated their own marketing drives. Interviews with selected regional co-ordinators elicited a mixture of praise and criticism. For example, Peter Green of Henry Cooke Lumsden & Co. (regional co-ordinators for Manchester, Derbyshire and Lancashire) thought the train to be a 'very good novel idea' but admitted that it had been very difficult to get professional advisers to spend valuable fee-earning time to attend. Those who had attended found the briefing valuable and interesting although it was found too technical by some. Subsequently the firm began its own marketing drive holding eleven seminars for the public in Preston, Blackpool, Rochdale, Manchester Central, Derby, Matlock and the Manchester region. At each seminar, names and addresses of those interested were taken. Martin Ellis, of Stock Beech & Co. of Bristol, echoed the novelty of the train and acknowledged the real difficulty of meeting their quota of guests partly because of the short notice. Their own marketing drive included the distribution of Dewe Rogerson's information packs to financial intermediaries in Avon, Somerset and Gloucestershire. In addition the firm wrote up to 1300 letters to intermediaries and held eight presentations. Derek McIntosh of Bell, Lawrie, MacGregor & Co. of Edinburgh also had difficulty in persuading intermediaries to attend so far in advance of the flotation. Invitation cards were followed up with persuasive phone calls and they did reach their quota. They found that the train was useful in encouraging those who had attended to join in later seminars and distribution activities. For the flotation, the company took space in the House of Fraser store in Princes Street and set up a BT stand with stockbrokers Wood McKenzie in order to reach the man in the street. It was remarkably successful. The stall attracted hundreds off the street, 75 per cent of whom had never before invested in equities.

For Belfast, the last intended destination, the train was left behind and a presentation was made in the Forum Hotel. Campbell Murray of Wm F. Coates & Co. explained their extensive marketing activity. They visited 24 towns, and called over 170 bank managers, 53 solicitors and 40 chartered accountants. They distributed the information packs and brokers' reports and organised a presentation in Londonderry. Many people with whom they made contact had never invested in shares and those most voracious for information turned out to be students of business studies from Belfast and Newton Abbey Technical college who had been set projects on British Telecom, but there was extensive and genuine investment interest from members of the general public.

The train therefore met its objective of the dissemination of information on a wide regional scale; it moved people along the learning curve, established a liaison between regional co-ordinators and financial intermediaries in their area and sparked off the biggest exercise in retail selling of any equity issue.

INDIVIDUAL SHAREHOLDER INCENTIVES

To convert interest into action is a complex process requiring the creation and interaction of sufficient enabling conditions with positive motivational forces. In the case of BT the essential enabling conditions were provided by the Government in a variety of forms, of which perhaps the most significant was the low minimum investment of £250 of which only £100 was payable upon application, the remaining sum to be paid in equal instalments in June 1985 and April 1986. The low level of investment combined with instalment purchase deepened the target audience and the potential investor field.

Ensuring that the offer would be within reach of a large proportion of the population was, however, merely one battle in a long campaign. Capturing and maintaining interest over a long period and avoiding the pitfall of boredom is a difficult enough task for any communications programme and peaking the publicity simultaneously with the offer merely adds another dimension of complexity. The idea of shareholder benefits is not new, in fact it is so well established that long league-tables of shareholder perks are frequently published in the press. Nevertheless, the offer of shareholder benefits at the time of flotation is new and supported the motivational factor required to turn interest in shares to investment in shares. After much debate and market research the Government chose to provide at its own expense special incentive schemes available for individual investors. First there was the telephone voucher scheme under which vouchers worth £18 each could be used to pay part of a telephone bill. The number of vouchers were to be directly linked to the level of investment and the length of time the shares were held (Table 3.6).

Table 3.6 *The telephone voucher scheme*

Total investment in £s	Number of shares purchased	Holding period Jul. 1985	Jan. 1986	Jul. 1986	Jan. 1987	Jul. 1987	Jan. 1988	Total number of vouchers
260	200	1						1
520	400	2						2
1040	800	2	2					4
1560	1200	2	2					4
2080	1600	2	2	2	2			8
3120	2400	2	2	2	2	2	2	12

As an alternative to the bill voucher scheme, individual investors were offered a share bonus of one free share for every ten purchased at the time of the flotation up to a value of £5000 and held for a full three years. MORI research, based on random samples of 2000 people indicated consistently that the majority of applicants would favour the bonus issue, and in the event the research proved correct with 64 per cent choosing the bonus share option.

The structure of the individual shareholder benefits as devised for the flotation increased the incentive for investors to hold their shares for a number of years. So far as non-employees who opted for the vouchers are concerned, because of the scaling-down the hold on them expires in January 1986.

PUBLIC ENQUIRY HANDLING

In what was effectively the largest commercial direct-mail operation in the world, British Telecom alerted its 20 million customers to its forthcoming stock market debut. Between July and September 1984 British Telecom customers received a flotation leaflet with their quarterly telephone bill, directing them to the British Telecom Share Information Office (BTSIO) at Bristol for further information. Simultaneously advertisements steered the public towards the BTSIO either by telephone on 0272 272 272 which was to become the most widely known telephone number of 1984, or by writing to PO Box 1, Bristol BS99 1BT, a most distinctive address.

Telephone enquiries were dealt with by Telecom Tan, British Telecom's own computerised answering service, operating seven days a week and equipped from the very beginning to deal with 50000 enquirers a week. An operator would take down names and addresses which would be entered into the computer and passed on to the appropriate regional mailing houses from which an information pack was sent directly to enquirers' homes. Postal enquiries and information packs were handled by Mail Marketing (Bristol) Ltd.

Table 3.7 *Schedule of information pack distribution*

Time	Information pack of:	Quantity
From 8 July to 31 August	Stock Exchange booklet, BT corporate brochure and letter	135 000
to 22 October	Stock Exchange booklet; BT corporate brochure, fact sheet and letter	750 000
29 October	Mailing of 'Action' phase fact sheet with letter	750 000
by 12 November	Mailing of Stock Exchange booklet, BT corporate brochure, new fact sheet and letter (to new applicants)	450 000
19 November	Mailing of 'mini' prospectus with letter and pre-paid envelope	1 280 000

Source: Dewe Rogerson

The objectives were to provide a focal point for all public enquiries concerning the flotation as well as to distribute information when it became available (Table 3.7). At the same time a central mailing list was built up of prospective investors to whom the 'mini' prospectus together with a pre-paid envelope could be sent. From the daily volume of enquiries, Bristol could give the marketing committee weekly feedback as to the level of public interest. For example, on 5 September 1984, Dewe Rogerson was able to report to the marketing committee that the daily number of enquiries was occasionally reaching 25000, a level that was necessary to achieve a total response of 1.5 million. A month later the current rate of response, if continued, would achieve between 1.25 and 1.4 million enquiries. Such information was welcome and reassuring news to the marketing committee. It backed up MORI's research findings on the scale of public interest; it made believable the media public relations campaign, and by indicating a strong potential demand from the retail sector reassured and finally boosted institutional interest as it became clear that demand would exceed supply—the most happy of all situations for a new issue to be in.

Equally significant, postal enquiries were to yield valuable feedback of a wide-ranging nature. Letters and questions reflected popular misunderstandings and ignorance. They identified what people actually wanted to know, or what worried them about the offer. Such qualitative information, where significant, could be acted upon in the questions and answers in later fact sheets, and in news stories to family finance journalists. A MORI survey of 100 people who had contacted the BTSIO by telephone between 17 and 18 September showed that enquirers were predominantly male, middle class and over 45 years of age. A quarter of those interviewed said they were certain to buy shares in BT, and a further 40 per cent said they were likely to buy. Most interestingly, of the sample 63 per cent said they would *not* seek further advice and this was good news for the Treasury, for those who finally filled in their 'mini' prospectus and sent it in the pre-paid envelope would by-pass the intermediaries and thus save the Government the 2 per cent commission. In fact slightly less than half of the applications were stamped by financial intermediaries, including banks and brokers.

Most enquiries from the public fell into a distinct pattern, so much so that 30 standard answers were devised. Enquirers resident abroad seeking information packs and prospectuses were advised that owing to legal restrictions such documents could only be provided to a UK address. A British citizen, non-resident in the UK, should contact a bank or solicitor. There were numerous letters from UK residents for advice on how to apply for shares if they were on holiday during the application period. Detailed explanations on the operation of the incentive schemes were provided, and advice on how to buy shares for grandchildren or wives. More generally, many enquiries revealed a deep ignorance amongst the general public. Many perceived the shares as a building society or bank deposit—a long-term investment which could be cashed in when the money was required irrespective of market conditions as the following questions indicate—

'Will I get a statement every month?'
'Why will the bill vouchers stop?'
'Why is the share price 50p when its nominal value is 25p?'

Many did not understand the nature of the dividend and preferred to call it the interest rate. A small sample will illustrate some of the questions in people's minds:

'I am 91 years of age so am on borrowed time, instead of paying by instalments will it be possible to pay in full on allotment?'

'At the present moment I do not have TV even so I rarely watch any programmes, therefore I would fail to see any advertisement concerning the application date.

> Secondly I do not purchase or read a daily newspaper, therefore once again I will not see any advertising about the application date. Is there a means by which I can be sent an application direct as soon as they become available?'

> 'If I was considering a minimum investment of £250 then I would pay a deposit in November, subsequent payments would be in 1985 and 86. This means with my first payment my money only buys £100 of shares. Not till all the money has been paid will I have the allocated £250 of shares. After paying the deposit share prices will rise which means with my second payment I will be buying less shares. Therefore my question is as follows: can I make one complete payment in November, instead, thus buying all my shares at once and not spreading payment?'

The BTSIO handled thousands of enquiries with standard answers where possible; more complex queries which defied them were transferred to the BTSIO in London.

The public response was phenomenal, building up from a daily average of around 5000 in mid-July to a peak of over 36 000 in November (Figure 3.12). The average daily rate of response to the 'account mailer' grew gradually after July and doubled following the start of the offer advertising campaign on 20 August to reach 11650 in the first week and increasing gradually to a peak of 19452

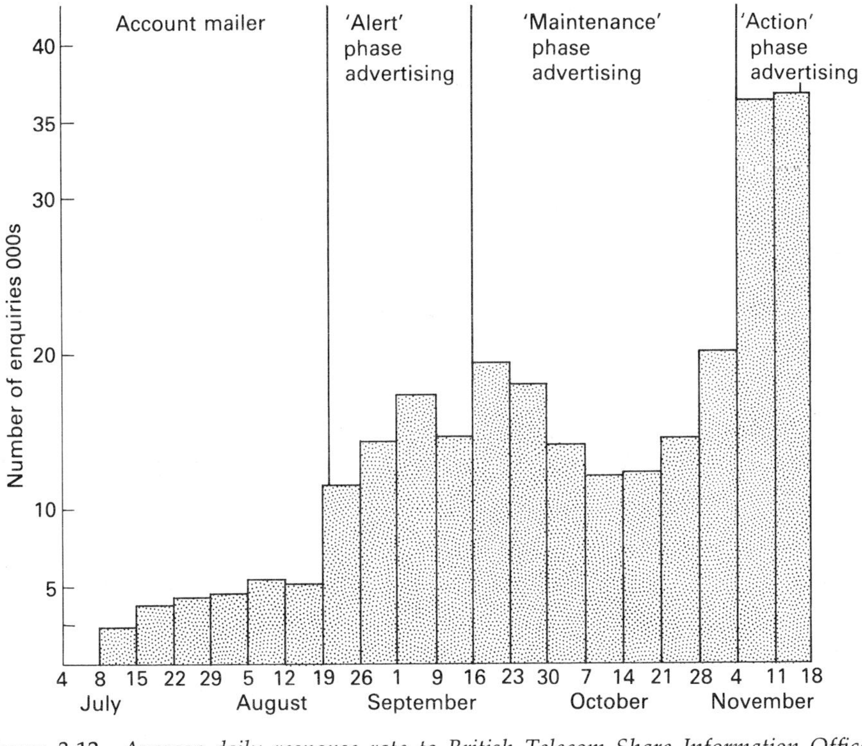

Figure 3.12 *Average daily response rate to British Telecom Share Information Office. Source: Dewe Rogerson.*

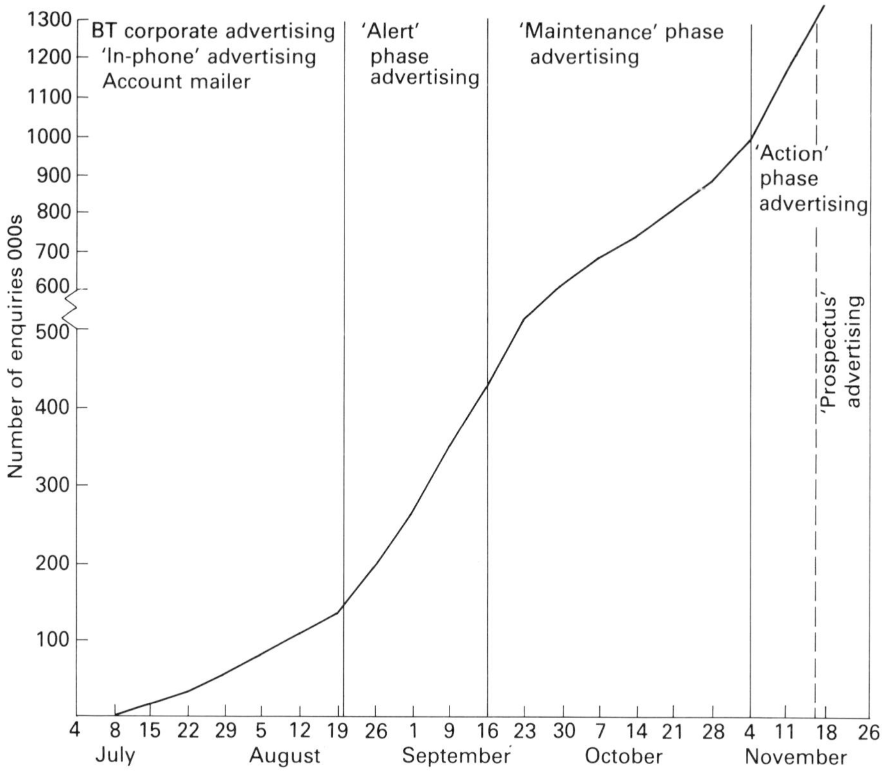

Figure 3.13 *Weekly cumulative total of enquiries to the British Telecom Share Information Office. Source: Dewe Rogerson.*

during the first week of the 'Maintenance' phase. The public response fluctuated with the level of advertising. A decline in advertising in the last week of September saw a fall in the level of daily enquiries to 18206 and the further decline in the first week of October was reinforced by the ending of the quarter's 'account mailer'. In the week ending 29 October interest revived, fuelled by favourable editorial comment following the launch of the Pathfinder prospectus. By 29 October nearly a million enquiries had been received (Figure 3.13). With the approach of the application date enquiries reached a crescendo and the number of telephone lines was increased to 150. On the weekend of 9–11 November 72587 enquiries were received of which 48000 were phone calls. By 14 November the total number of public enquiries to the British Telecom Share Information Office was some 1380000, of which 650000 were telephone calls and 730000 were letters.

On 19 November the BTSIO co-ordinated the postage of approximately 1280000 'mini' prospectuses with covering letters. The office and direct-mailing operation fulfilled a multiplicity of objectives. Of prime importance was the supplying of essential information on the timing, incentives and general features of the offer; conveying the character of British Telecom as a dynamic investment vehicle, and welcoming the potential investor to an exciting new adventure. In all this the operation succeeded remarkably well. The enthusiasm of a now well-informed public was translated into envelopes containing cash—among them even a cheque for £30 000 before the offer had opened. In the end, of the 1.3 million who enquired, 800 000 filled in their application form from the 'mini' prospectus and sent it with a cheque—a remarkable conversion rate of 61 per cent.

Chapter 4

Advertising Strategy and Execution

As early as November 1983 the marketing committee accepted what was to be the central strategy for meeting the Government's objectives and assuring the successful debut of British Telecom on the Stock Exchange in London. The key to success was seen to lie in completing two distinct stages, the first of which was the mass marketing of British Telecom as a company. Such a campaign had been mounted from 1 November 1983 under the slogan 'The Power Behind the Button'. It was devised by BT's corporate advertising agency, Dorland, and, as we shall see, was to be a vital pre-condition for the share-offer campaign. This, the second stage, was an even more demanding exercise—to promote the attractiveness of British Telecom shares to the general public. The rationale for this approach was compelling given the Government's declared wish to increase share ownership in privatised companies as a first step towards the achievement of a share-owning democracy against what was truly a bleak backcloth. Over the last twenty years there had been a fundamental change in the share-ownership profile, away from the private shareholder to the institutional investor. In 1963 private individuals held well over half the value of UK equities; by 1981 their share had fallen to around 28 per cent. Of the 43 million adults in the UK, only some 2.5 million owned shares or unit trusts according to an NOP survey of 1983 (Table 4.1) and this represented a very small potential investor audience given the scale of the forthcoming flotation. Moreover, of the 1.6 million estimated to own stocks and shares some 600 000 were employee–shareholders whose stake and experience was confined to a single company—their employers. Furthermore, based on a host of previous shareholder studies, the response rate from the existing shareholder base was expected to be small. Of the one million independent private investors over half are over 65 years of age and passive, in that nearly a half had held their shares and units for over five years. It was felt therefore that the older investors were not going to sell existing holdings in order to buy British Telecom shares. Instead they would

Table 4.1 *Profile of holders of financial assets, 1983*

Category of holding	Percentage of GB adult population	Holders in 000s
Stocks and shares	3.8	1558
Unit trusts	2.5	1025
Life assurance	52.2	21402
Bank deposit account	34.5	14145
Building society:		
— Ordinary share account	47.7	19557
— Term shares and higher interest accounts	5.6	2296
Premium Bonds	32.1	13161
National Savings Bank	11.4	4674
National Savings Certificates	6.1	2501
Savings at home	6.3	2583
No savings	19.9	8159

Based on UK Adult population (16 years or more) of 43 million
Source: NOP Financial Research Survey, Great Britain, October 1982–March 1983

have to increase their investment, a requirement which might be difficult to meet given the constraints of retirement. And even a high response rate of 50 per cent from this core target would achieve a number little more than double the number of existing shareholders in Shell or ICI with 340000 shareholders or 390000 respectively. On these grounds alone it was obvious that the target group had to be dramatically extended, not only in order to ensure the largest number of private shareholders for British Telecom, but also to avoid the disappointments associated with previous privatisations where the initial shareholder response proved to be shortlived. For example, there had been 158000 successful applications for British Aerospace shares and 125000 for Jaguar. Four months after the Aerospace flotation only 27000 shareholders remained, whilst Jaguar's shareholder register had shrunk to some 50000 in the space of a month. These considerations alone required an enlarged target audience, but how large should it be?

At no time, either at the beginning of the strategic planning nor during its execution, did the Government or its representatives at the Department of Trade and Industry or the Treasury commit themselves to a firm shareholder target for British Telecom. The Government's target therefore was never made clear in terms of actual numbers. Was it to be one million, or perhaps five million? What was clear was the hope that it would exceed existing numbers in individual companies (Table 4.2).

Given these circumstances, Dewe Rogerson from the outset of the strategic planning exercise, established a target audience for the retail marketing campaign of some 20 to 30 million adults, based on all residential British Telecom telephone customers, i.e. some 14.5 million households or 30 million adults. In this group those with savings of over £1000 and those households

Table 4.2 *Share ownership in selected British Companies, 1983*

Company	Number of shareholders in thousands	Market capitalisation £million	Rank in *Times* 1000
ICI	390	2954	5
Marks and Spencer	241	2654	6
BP	249	7562	1
Shell	340	6248	3
Imperial Group	150	658	35
Cable and Wireless	28	1188	20
Britoil	39	1104	21
British Aerospace	25	423	69

Source: Annual report and accounts

with an annual income exceeding £7500 would be specially targeted. By definition, therefore, the target audience was one generally inexperienced and uninformed of the equity market, dissociated from professional financial advisers, and ignorant of the methods of buying and selling shares. Therefore the first step in the retail marketing campaign must be to educate and inform the general public. Furthermore the campaign had to break through the barriers of perceptions and inertia surrounding share ownership. As the figures clearly demonstrate, share ownership was in practice and perceived to be the preserve of the well-to-do. It was for 'others'; it was not for the ordinary person safely locked into the familiar building society or bank account, National Savings Certificates or, at the most adventurous, Premium Bonds. To secure participation the campaign had to attract the interest of half the eligible adult population.

The level of interest was crucially dependent on two major factors. One would be the realisation by the target audience that they were being invited to participate, indeed that this offer was for *them* and relevant for any savings they might have. A second would be to transport them from such awareness to participation and this required a positive motivation to buy British Telecom shares rather than shares in general. Provided that the Government could deliver the right balance of security, price and terms of payment, the communicators could deliver the shareholders with a suitable and large-scale marketing campaign.

In mid-February 1984 Dewe Rogerson submitted a detailed paper to the marketing committee. In addition to a mass publicity campaign for the share offer aimed at twenty million BT customers it proposed a tailored communications programme aimed at financial intermediaries, including accountants, solicitors, insurance brokers, financial consultants and bank managers. Their significance lay in two contexts. As individuals they were potentially the most receptive audience to respond to the share-ownership call: they themselves were likely to take up their full allotment. Secondly, if they were properly briefed and motivated, their role in helping to explain the offer and its attractions would generate new business and widen the market response. Their role in the distribution of background information on British Telecom and the offer would be an invaluable addition to the traditional distribution network confined to the City and existing investors.

The paper furthermore sketched in outline the communication messages which would be required and set out their rationale in a series of logical propositions. First, in order for the public to be interested in British Telecom the audience needed information on British Telecom and sufficient understanding to take a view on BT's potential for growth and profit, hence the importance of the existing corporate campaign. Upon the assumption that the public had acquired a favourable image of BT, the next step—the decision to invest in British Telecom—required the investor audience not only to know that shares in British Telecom were to be sold but to know the mechanics of how and where to buy shares. They needed not only to regard British Telecom as a superior home for their savings but to feel that share purchase was simple, easy and as convenient as going to the bank or post office. To meet these requirements the communications programme required the two distinct messages—the corporate campaign concentrating on British Telecom the company, already being conducted by BT, the other the campaign to promote the share offer by the Government. Both sets of messages were to be kept distinct in terms of content and origin—two communication programmes would have to be run consecutively. The corporate campaign and the offer campaign had to be kept separate for a number of sound reasons—to comply with the Companies Acts and advertising codes which would not permit their amalgamation, to obey ethical standards and avoid accusations of share pushing, to avoid confusing customer and investor interest, and lastly for plain old-fashioned practicality the offer campaign needed to be flexible, and therefore a separate campaign was the solution.

That British Telecom's corporate campaign had already achieved substantial improvements in BT's various image profiles was confirmed by MORI research, but their research also demonstrated perceived image weaknesses. The public had an inadequate understanding of British Telecom in four crucial areas. Whilst everyone linked British Telecom with telephone services and hence a domestic utility image, in fact British Telecom's operations were highly diversified. Few of the public had any conception that the everyday object, the telephone, when combined with sophisticated technology, was fast producing a whole new business commonly known as information technology. Moreover, if the public thought about British Telecom's management at all, they had a low perception of its quality and strength. Finally, research indicated some concern over potential constraints on profitability arising from the Government's influence on British Telecom's pricing policy and the unions' anti-privatisation stance. The existing corporate campaign of British Telecom evolved to meet these research findings and the requirements of the offer programme. The modified corporate campaign needed to address the public's ignorance of British Telecom's scope and diversity of services, stressing their range, scale and past growth as well as displaying the multiple applications of BT's existing operations. Given appropriate execution, the campaign could demonstrate British Telecom's prowess and thereby shift perceptions of British Telecom from static public utility to high-tech growth company. At the same time such demonstrations of BT's expertise would indicate outstanding growth potential and shift implicitly negative regard for BT's management to positive views, for having achieved so much and for planning for the future. The public's ignorance of information technology needed to be addressed to reinforce perceptions on the range, scale and growth opportunities available to BT, and the impressive scale, diversity and continuity of BT's investment programme required portrayal with an emphasis on its contribution to the economy and public welfare. This aspect would further reinforce positive views of British Telecom's management, demonstrate the compatibility of profit and competitive pricing and underline the achieved and potential improvements to customer services. Finally, to counteract fears of the effects of privatisation on the provision of essentially social services British Telecom's commitment to maintaining and improving such services had to be stressed.

The offer campaign needed to run for several months given its novelty and the complexity of the messages. The messages needed to be clear, simple and repetitive. First and foremost it was essential to inform the target audience that British Telecom was to be privatised and that shares were to be sold to all those who wished to apply. A lengthy campaign was required in order to give advance warning of the forthcoming event and help an unsophisticated public to become accustomed to the idea of buying shares and ready to buy when the offer was made. Furthermore, information was crucial to the investment decision; those interested needed to know not only that information would be available but where to obtain information on British Telecom and share buying. The public needed to be assured that they would be kept informed and that when the time came, it would be simple to understand, and simple to apply. They had to be made aware of the availability of the prospectus and of free financial advice. When the crucial time of application arrived they had to be assured of a clear and simple basis for decision and action.

All parties were confident that the gap in the public's knowledge and understanding could be closed with an integrated share offer marketing campaign. However, the creative execution of the themes and messages required further work. BT's audit of the existing corporate campaign, due for completion at the end of February 1984, would provide the basis for a detailed creative brief shifting the emphasis of the remaining stages of the corporate campaign. Given its novelty and complexity, qualitative and quantitative research into beliefs, attitudes and knowledge would have to be undertaken to guide the creation and execution of the offer campaign. What was clear, as early as February, was not only the unusual duration of the proposed campaign but the recognition that the campaign be divided into phases reflecting different aims, content and intensity of publicity. Altogether four distinct phases for the share offer campaign were conceived by Dewe Rogerson—'Alert', 'Maintenance' and 'Action' culminating in the 'Prospectus' phase.

Towards the end of March 1984 Dewe Rogerson submitted to the marketing committee a progress report together with plans and costings for an integrated share offer communication programme and sought specific decisions on budget size and allocation, together with authorisation for specific media schedules and production of literature and advertisements. Substantial progress had already been made. An audit of existing MORI research was complete, the criteria influencing the investment decision had been identified, new areas for research had been isolated and a full research proposal to monitor public awareness, perception and motivation would be presented to the committee at the end of the month.

The national offer campaign was essential not only to mobilise private investors but also to demonstrate to the financial institutions—the traditional buyers of new issues—that the Government was committed to a public marketing campaign of unprecedented scale designed to sell at least one quarter of the shares to individual shareholders. Such public commitment on the one hand could reassure the reluctant institutions that they alone would not be expected to clear the new issue and on the other hand, by implication, would lead them to the conclusion that the issue price would be attractive. This was absolutely essential, for in the spring of 1984 the financial institutions expressed reluctance to buy and underwrite British Telecom stemming from the sheer scale of the issue and concern over the price.

The marketing key therefore was to create substantial demand in a situation of excess supply—to create an image of scarcity for what was plentiful and to imply an attractive price against fears of an unattractive price. The catalyst would be the demonstration of the Government's strong commitment to wider share ownership conveyed by a huge publicity campaign from which the institutions must conclude that their perceived problems of underwriting and buying shares would be diminished by at least the extent of the public response. Clearly, if the price was to be right for a hitherto uninitiated public, it would be right for the sophisticated institutions. The same line of reasoning would apply to the existing core investor group and the financial intermediaries. This whole edifice of expectations and requirements, capable of reconciling Government, the public and the City, was crucially dependent on the terms of the British Telecom offer. 'Without a sufficiently attractive offer, the marketing will fail, the wide share ownership objective will not be achieved, and British Telecom and the Government will have a public relations problem of a major and basic order' warned Dewe Rogerson.

THE CREATIVE EXECUTION OF THE BRITISH TELECOM 'POWER BEHIND THE BUTTON' CORPORATE CAMPAIGN

In November 1983 British Telecom had launched its corporate advertising campaign, 'The Power Behind the Button'. Devised by Dorland Advertising, British Telecom's corporate advertising agency, the campaign concentrated on television in order to reach a mass audience and focused on British Telecom the company. The objectives were to raise selectively awareness of what BT is and does. As the advertisement demonstrated, BT was far more than just a basic telephone service company or public utility: British Telecom's strengths were the wide range and scale of its services, its technical sophistication, its research depth and its commanding position to take advantage of new growth opportunities. It was a company vital to the community both domestically and internationally. The different story-lines held together by the use of unified visuals, music, tone of voice and repetition, created a new image of British Telecom. Gone was the hum-drum telephone company—in was the hi-tech company of the future. The campaign, according to MEAL figures costing £16 million of which nearly 90 per cent was accounted for by television, obtained a high profile for British Telecom and was incidentally an effective voice to counteract the unions' anti-privatisation stance. With over 20 individual advertisements the corporate campaign built up the high-tech image, layer by layer, moving across the whole range of services from the domestic telephone to satellites. Then, from Spring 1984, the advertisements began to embrace key concepts dear to any potential investor, namely BT's size, strength, range and scale of services, commitment to investment, sophisticated technology and the growth opportunities represented by the information revolution. This helped to lay the foundations for the specifically designed offer advertising that was to follow.

At the start of the corporate campaign the advertisements stressed that what the audience knew and saw of British Telecom was merely the tip of the iceberg. British Telecom had hidden depths and the campaign interlocked the mundane and futuristic aspects of BT's operations. At one end of the spectrum, BT was shown to lead the field in laser-power communications, their threads of glass making possible the carriage of 2000 calls simultaneously. BT technology carried what we see and hear on our television and radios and from the automated network system and earth satellites the UK has the most extensive direct overseas dialling system in the world. Such applications of the present technology 'Keeps us and the future in touch'.

The second phase of corporate advertisements specifically addressed the key anxieties arising from the effects of privatisation. A series of advertisements emphasised the social role of the telephone in a push-button world, and added the reassurance that British Telecom remained dedicated to improving and maintaining the free emergency services provided under 999. These services to call aid from the police, fire, ambulance and air sea rescue services were used 35000 times a day to help people, and BT remained committed to them. Just as important was the provision of rural telephone kiosks for remote communities and stranded motorists, used in emergencies and the normal interchange between friends and relations. This service too was safe for the future and would even be improved, BT reassured its audience.

The third phase of the corporate campaign dwelt on the technological revolution, the scale of operation and investment, and the extent of research and development. It concentrated on growth and opportunities whilst at the same time stressing the importance of BT in ordinary life. Starting from acknowledged home ground, BT reminded the audience of the variety of telephones now on offer, giving choice of colour, size and shape and demonstrated how sophisticated telephones had become—from intelligent phones with memories for frequently dialled numbers to phones that took phone cards instead of cash. There were incredibly sophisticated telephones for the office. For example, the City Business System was a miracle of technology used by money dealers in the City of London and overseas which by a touch on the computer screen linked dealers across the world. Such a system was helping London to stay at the heart of the world's financial markets. The series concluded with what was to become a symbol of the flotation campaign, a satellite dish aerial at BT's Goonhilly Downs earth station which underpinned national and international communications, in which BT was a world leader.

Then the spotlight was directed on the wizardry of the National Networks divison which among other things linked the banks' hole-in-the-wall automatic teller machines with the banks' computers, made advance holiday bookings possible, and made business communications more effective. Once again, BT was shown to be shaping the future through accelerated technological change and investment. Joanna Lumley, amongst other personalities, demonstrated the advantages of System X (the new computer-controlled digital telephone exchange) over the old slow and mechanical exchanges. Martlesham, BT's research and development centre in Suffolk, was already testing person-to-person visual communications, something taken for granted by post-war science fiction writers but so far not in general use. Powerful visuals were intertwined with messages of a speed beyond comprehension, the promise of quicker and quieter connections and the prospect of lower operating costs. It was shown how a multinational oil company's North Sea rig relied on British Telecom for its worldwide data and communications links. Finally, to bring home the message that this technology really was available and in use, the final batch of advertisements concentrated on the Olympic Games, blessed with simultaneous transmissions, giving live presentation on the screens of over twenty European countries via BT's network and earth stations. Each of those advertisements began with the dish aerial, the technology that made simultaneous transmission possible, the symbol for a mountain of sophisticated technology, encapsulating present and future, which ultimately became the branding device of the flotation publicity. The campaign was a powerful visual demonstration of BT's areas of operations, its crucial presence in the community, its use of sophisticated technology and its growth opportunities. As each of the advertisements stressed, 'BT is the Power Behind the Button'.

In June 1984 the corporate campaign was joined by a call-stimulation campaign, designed to increase telephone calls and the company's revenue. The touch was light and humorous but enormously successful in giving BT a higher public profile with advertisements humorously portraying famous or legendary characters—Quasimodo, General Custer, Tarzan and King Neptune. Each episode was interrupted by a well-timed telephone call which gave the nation a new catch-phrase 'It's for you-hoo!'

Throughout the duration of British Telecom's corporate campaign, MORI's tracking studies monitored the impact upon the public and concluded: 'There is no doubt that British Telecom's overall advertising, including the corporate campaign, has had a significant impact upon the public image of the corporation . . . 11 out of 15 image ratings which we have monitored monthly have gone up month by month'. Thus, the crucial rating of hi-tech versus public utility measure rose by 13 per cent, a key perception in the acceptability of the shares in the market-place.

THE OFFER ADVERTISING PROGRAMME

The 'Offer' advertising programme had four distinct phases, the content of which will be examined in turn. Each had different objectives which would converge in the ultimate end of a widely-distributed and widely-understood prospectus reaching out to a well informed and sympathetic audience.

(a) The principal objectives of the 'Alert' phase were to put across three basic messages: that British Telecom was to be privatised in the autumn, that the Government was inviting members of the public to apply for British Telecom shares, and that anyone interested could phone in or write for further information. In response the British Telecom Share Information Office would send out the British Telecom corporate brochure describing BT the product, and the Stock Exchange booklet on share ownership. The essential hall-marks of the 'Alert' phase advertising would thus reinforce on a much larger scale of communication the 'account mailer' invitation to apply for further information and provide a considerable impetus to the direct response campaign. The novelty of the invitation combined with the favourable image of BT, created by the 'Power Behind the

Button' corporate campaign, would transform awareness into interest. The information packs and high-profile advertising together would create acceptance and familiarity with the invitation call; it would normalise the idea of buying BT shares amongst the public.

(b) The 'Maintenance' phase was designed in part to reinforce the objectives of the 'Alert' phase—to build awareness of the invitation to buy and to encourage requests for further information, thereby supporting the direct-mailing programme and extending the market. At the same time, additional information on the specific offer could be transmitted. By virtue of the continuity and high profile of the several campaigns on television and in the press, combined with the anticipated media discussion and endorsement of the sale, the idea of doing something wholly uncharacteristic with their savings and the perception that owning British Telecom shares would be a good thing, would be instilled across the nation.

(c) The 'Action' phase, as its name suggests, was to represent a burst of advertising on all available media—television, radio, newspapers, magazines and posters. Here too there were three objectives. How to get the prospectus to those who had not yet responded, to encourage applications and to explain what was going to happen and to instil urgency and get potential investors poised and ready with cash to invest, for the offer was now imminent.

(d) Finally, in common with traditional offers for sale, there was the 'Prospectus' advertising phase. To comply with the Telecommunications Act 1984 the full prospectus was to be published in four national newspapers, the *Financial Times, The Times,* the *Daily Telegraph* and the *Guardian.* In addition to the statutory publicity surrounding the full prospectus it was proposed, given the extensive target audience, to send an abridged version of the prospectus (to be known as the 'mini' prospectus) to all enquirers on the mailing list, and furthermore to advertise it in the popular press and selected regional newspapers. Every ancillary advertisement during this phase would draw attention to the availability of the prospectus in order to merchandise it and equally important in order to comply with the Independent Broadcasting Authority's rules.

'Alert' phase, 20 August to 16 September 1984

The date 20 August marked the end of British Telecom's corporate advertising campaign and the start of the four-stage offer advertising campaign costing £10.5 million. The offer advertising campaign was carried out by Dorland Advertising and involved 25 000 work hours of preparation and over 100 people working in conjunction with Dewe Rogerson.

Among the first advertisements of the four-week long campaign was a notably distinct one in terms of its selectivity of audience, media and message. Whole-page advertisements appeared during the first three days of the campaign in the *Financial Times* and *The Times* and specialist financial press announcing to the City and financial intermediaries that the British Telecom share issue was being designed to attract the widest public response, based on a minimum investment level of 'about £250' with instalment payments over 15 months and a special package of benefits for individual share owners. In essence this announcement sought to demonstrate to the traditional participants of a new issue that the British Telecom flotation, supported by a widespread retail marketing campaign, really was being extended beyond the traditional investor base to the first-time buyer. In effect, it offered reassurance to the City, concerned with the prospect of a large-scale funding burden, that everything possible would be done to support the call for mass ownership of British Telecom shares.

Simultaneously the mass public advertisement campaign began, using a media mix of television, national and regional newspapers and radio in order to achieve an immediate high profile for the British Telecom share sale. Its specific objectives were to communicate that the offer was for everyone, and to attract a response from which a mailing list could be compiled. The latter would serve to assess the degree of public interest and the effectiveness of the campaign, and facilitate the distribution of the 'mini' prospectus to be published in November. Thus each and every advertisement encouraged a response, either in the form of a telephone call or by clipping and posting the coupon to the British Telecom Share Information Office.

A major task of the campaign was to ensure that the call to share ownership became regarded as relevant, and that the audience identified with the message. This was achieved by a winning formula, featuring a representative sample of the population, balanced by age, sex and occupation. For example, a housewife, garage manager, truck driver, supermarket manager, roof tiler, tractor driver, restaurant manager, secretary, computer operator and typist appeared in groups known as composite advertisements under the slogan, 'Now everyone can share in British Telecom's future' (Figure 4.1). Other advertisements reinforced the identification process by featuring the same individuals alone with the object of linking occupational class with the idea of sharing in British Telecom's future (Figure 4.2). Everything possible was done to support the widespread call for ownership of BT shares, including the assurances that 'You will have the opportunity to be an owner of a company', and that applying for shares would be easy, and that advice was available from banks and financial intermediaries. The other crucial message was that the minimum investment was to be about £250 of which less than half was payable immediately. The appeal of special benefits was thrown in, but to add a note of caution (and perhaps, to add a spice of excitement calculated to appeal to the

Figure 4.1 *Composite press advertisement.*

gambling instinct) the warning was added, 'You will appreciate that the market value of any share can go down as well as up'. Overall, in design and style, the approach of this series of press advertisements was informal and popular.

The 'Alert' campaign featured advertisements on commercial radio with a more formal tone of voice conveying the simple messages, that a prospectus

Figure 4.2 *Press advertisement.*

would be published in November and that 'You don't have to be a big City investor. Anyone who wants to can apply for British Telecom shares'. To reserve a copy of the prospectus and receive further information on British Telecom and share-ownership listeners were directed to newspaper advertisements for the coupons or the telephone number 0272 272 272.

The medium with the largest coverage of all target groups, was, of course, television. Two separate 60-second commercials were prepared and transmitted. Both were designed to stress the official Parliamentary sanction of this unusual exercise. Both combined extreme formality, exemplified by voice-over, with the solemnity of the opening scene and message. The first advertisement opened with a view of the Chamber of the House of Commons against the spoken preamble of the Telecommunications Act, April 1984, and some of its provisions, namely that 'on the twelfth of April, 1984, Parliament made it possible for the people of the United Kingdom to own British Telecom shares'. It soon moved on to the popular appeal, 'That means not just big city investors but everybody who wants to' and ended with an invitation to 'Reserve your copy *now*'. It was the kaleidoscopic visual element of the commercial that democratised the share-ownership call, including as it did several snapshots of British life. The dawn of a new day in the countryside was matched with the urban milkman's round, the sight of shops opening up, of children going to school, commuters alighting from a train . . . From this the advertisement moved to a visualisation of the dominance of communication in our lives, illustrated by the purchase of a newspaper, the use of a cardphone, people meeting each other and finally rural and urban scenes—all branded with the concluding emblem of the Goonhilly satellite aerial against a frieze of representative citizens (Figure 4.3).

The second television commercial, shown in September, again used a formal tone commensurate with the gravity of Parliament and the Telecommunications Act 1984. Formality was once again counter-balanced by scenes of activity and different landscapes, (Figure 4.3) and branded by the slogan 'Find out about sharing British Telecom's future'. Between them, the press, radio and television advertising of the 'Alert' phase succeeded in attracting 200 000 enquiries to the British Telecom Share Information Office.

'Maintenance' phase 17 September to 4 November

The seven-week 'Maintenance' phase campaign featured a direct response element, with coupons in the print media and the British Telecom Share Information Office telephone number on press advertisements. Posters and television commercials invited requests for information and later went on to urge a booking of the prospectus. The combination of an intricate media mix, and intense frequency of advertisements created a ubiquitous presence for British Telecom. Reinforced by the media public relations campaign, the advertising campaign effectively made British Telecom and the share offer a feature in all media and combined synergistically to make British Telecom and its flotation a national event. British Telecom assailed the eyes and ears of the population from all directions and became a universal topic of conversation.

The 'Maintenance' phase of the campaign had three distinct objectives. Firstly, having awoken the interest of a large number of the general public, it

Figure 4.3 *Stills from a TV advertisement.*

TELECOM

TELECOM

became important to continue to stimulate public debate amongst ordinary people. The strategy was designed to keep British Telecom's flotation a talking point up until the prospectus was available. Secondly, there was a need to impart certain specific information about the share issue to a public who were unfamiliar with the ways and dealings of the City. Thirdly, as with all four phases of the campaign, it was vital to elicit response from the general public so that the all-important mailing list could be prepared. The marketing committee recognised the need to put a prospectus directly into the hands of as many interested purchasers as possible on the day it was to be published.

The advertising was a mixture of television, national press, magazines and posters. On television six 'vox-pop' commercials were made. This is a technique which allows ordinary members of the public to express in their own words what they think about a topic. Because of the legal constraints which prevented any hyping of the British Telecom flotation, the advertising allowed the people being interviewed to put their own points in the form of questions. For example, one lady said: 'I don't read any financial newspapers, how would an ordinary person like me find out about buying shares?' At the end of the commercial, the British Telecom Share Information Office telephone number was given so that people could telephone for the information pack and find out all the answers to their questions. The national press made a direct link with the television advertising by featuring in the press advertisements people who had appeared in the television commercials, and it was possible in the press advertising to say far more about the share offer than was possible on television. The newspaper advertisement in Figure 4.4 tackled head-on the twin information gaps—the public's unfamiliarity with share buying and their lack of information about British Telecom itself. By filling in the coupon, or by telephoning, which involved no commitment or obligation, respondents would receive the information pack containing the essential background to British Telecom, information on the mechanics of buying shares together with a leaflet explaining the special benefits attached to ownership of British Telecom shares. In November a specially designed prospectus in plain English would be sent outlining the terms of the offer and containing an application form. The newspaper advertisement in Figure 4.5 outlined the specific benefits of the British Telecom share offer, explained the voucher system and contained the customary coupon, branded as usual with the flotation symbol—the dish aerial and its attendant admirers. The newspaper advertisement in Figure 4.6 once again emphasised the share benefits available but focused on two major issues, the special instalment payments and, what was to be a key item in the 'Action' and 'Prospectus' phase, the time constraint on applications. The public would have only eight days in which to apply.

Magazines were included in the media schedule at this time—television programme magazines, Sunday supplements, general interest publications and women's magazines. Because of the more leisurely reading environment

Buying British Telecom shares.

Suppose I answer this ad today; what happens next and when?

"I haven't really got the faintest idea how I'd actually physically go about acquiring a share."

PETER SIMPSON

"Where is it possible to get the information for British Telecom shares?"

GEORGE HIGGS

If you've never bought shares before, your first reaction, once you start, could be one of surprise.

Because it's a whole lot easier than you might think.

In principle, buying shares is just like anything else. You hand over the money; the owner hands over the goods.

But before you do anything else, you examine what's on offer.

First, the information pack.

We have put together an information pack, specially for this sale.

It gives you a lot of background data about British Telecom – the networks, the satellites, the optical fibres and so on. All the 'power behind the button'.

You get the latest Stock Exchange booklet about buying, selling and owning shares. There are two main returns that can come from this: dividend income, and growth in the value of the shares, although neither can ever be guaranteed.

You also get a question-and-answer leaflet that tells you about personal founder shareholder benefits – cheaper telephone bills for up to three years, or a free bonus share issue for those who keep their shares.

In November, a prospectus.

A simple document will be published, written in plain English. Where an information pack gives you outlines, this document will set out what is on offer and includes an application form. It will tell you more about the latest figures and activities of the business and includes a forecast of profit and dividend.

It will tell you what the shares cost, and how you pay for them.

Suppose you invest £1,000. You would pay £400 when you apply in November, £300 in summer 1985, and the remaining £300 in early 1986.

No obligation, no commitment.

Replying to this ad means you are ordering your information pack and booking your prospectus. Nothing more.

Ring 0 272 272 272 any time of the day or night.

Or return the coupon to P.O. Box 1, Bristol BS99 1BT.

Find out more about sharing British Telecom's future.

Please send me details about British Telecom and share ownership.

MR/MRS/MISS ____________________ BLOCK CAPITALS

ADDRESS ____________________

____

____

POSTCODE ____________________

ISSUED BY KLEINWORT BENSON LIMITED ON BEHALF OF H.M. GOVERNMENT

Figure 4.4 *Newspaper advertisement.*

that magazines enjoy, the copy was written in a more relaxed style with headlines more appropriate to the environment (Figure 4.7), for example, the Sunday magazines featured a headline 'The Sunday morning guide to owning British Telecom shares'. The inclusion of women's magazines marked the importance the marketing committee attached to the joint decision-making process of smaller investors. One particular creative treatment appeared as the

British Telecom:

The benefits for founder shareholders.

"Will I get a telephone discount?"
RICHARD MADDOX

"Would it make any difference to my telephone bill?"
JILL BLOUNT

The main returns from investing in shares are dividend income and growth in share value. These, of course, can never be guaranteed.

But with British Telecom shares there are extra benefits limited to founder-shareholders and these are guaranteed.

Cheaper phone bills.

The first of these is the bill voucher. Each voucher is worth £18 against your telephone bill.
The number of vouchers you receive depends on how much you invest.

For example, if you invest £1,000 you will receive four vouchers; two in July 1985 and two more in January 1986.

The scheme lasts up to three years. In that time, it can produce a benefit worth up to £216, free of income tax, over and above any dividends if you keep your shares.

These vouchers are guaranteed on any investment up to £3,000.

Alternative: free bonus shares.

Instead of vouchers you may choose to receive a free bonus of 1 share for every 10 of your shares that you continue to hold after three years.

The bonus share issue will also be free of income tax. It will be necessary to state at the time of application which benefit you are opting for.

VOUCHERS: HOW MANY AND WHEN?

VALUE OF SHARES PURCHASED (Approx)	JULY 1985	JANUARY 1986	JULY 1986	JANUARY 1987	JULY 1987	JANUARY 1988	TOTAL NUMBER OF VOUCHERS
£250	1						1
£500	2						2
£1,000	2	2					4
£2,000	2	2	2	2			8
£3,000 OR MORE	2	2	2	2	2	2	12

Shares can, of course, go down in price as well as up: there can be no guarantee about this. But the bonus is guaranteed – for the whole of your investment up to £5,000.

Interested? Ring 0 272 272 272 any time of the day or night, or return this coupon to P.O. Box 1, Bristol BS99 1BT.

Find out more about sharing British Telecom's future.

Please send me details about British Telecom and share ownership.

MR/MRS/MISS ____________________

(BLOCK CAPITALS PLEASE)

ADDRESS ____________________

POSTCODE ____________________

ISSUED BY KLEINWORT, BENSON LIMITED ON BEHALF OF H.M. GOVERNMENT.

Figure 4.5 *Newspaper advertisement.*

British Telecom November share offer

The benefits of applying for British Telecom shares in the eight key days.

Time is always short when new shares are being offered. With British Telecom you will have eight days to get your application in.

The advantage is that if you apply for shares in these eight days you will be in line for a choice of benefits that will not be available to anyone else.

These are personal benefits, over and above all dividends; they are free of income tax, and for most people also free of Capital Gains Tax.

They are limited to founder shareholders only. You say which benefit you want when you apply.

Vouchers for cheaper telephone bills.

Each voucher is worth £18. The number of vouchers you get depends on how much you invest and how long you keep your shares.

Maximum discount is £216, with twelve vouchers. For this you need to invest £3000 or more, and you have to keep the qualifying shares for three years.

This scheme is not an alternative to dividends; it is additional. Unlike dividends it is guaranteed and it is free of income tax.

You can get vouchers only by applying for British Telecom shares in the November offer. They will not be available at any other time.

Free share bonus.

Instead of the telephone bill discount, you may choose a bonus of one free share for every ten of your shares that you still hold after three years.

This will apply to the first £5000 of your investment – a higher benefit ceiling than for vouchers – and is additional to any dividends.

The share bonus is free of income tax and limited to people who apply for their shares in the November offer. Nobody else will be entitled to it.

Special payment arrangements.

For example, if you invest £500, you would pay only £200 with your application; £150 in summer 1985; and the remaining £150 would not be due until early 1986.

The same principle applies right down to the minimum investment of about £250.

The price of any share can, of course, go down as well as up; but as soon as your initial payment is accepted you will have your share in one of Britain's leading companies.

You can put your name on the list for a prospectus here and now, by replying to this advertisement. This does not put you under any obligation.

Ring 0 272 272 272 at any time of the day or night – or return this coupon to P.O. Box 1, Bristol BS99 1BT.

Find out more about sharing British Telecom's future.

Please send me details about British Telecom and share ownership.

MR/MRS/MISS ______________________ (BLOCK CAPITALS PLEASE)

ADDRESS ______________________

____

POSTCODE ______________________

ISSUED BY KLEINWORT, BENSON LIMITED ON BEHALF OF H.M. GOVERNMENT.

Figure 4.6 *Newspaper advertisement.*

The armchair g
British Tele

If you already own ordinary shares you belong to a relatively small group of people. Fewer than two million individuals in all Britain own shares.

In November of this year, however, the sale of shares in British Telecom will give millions more an easy opportunity to participate in the fortunes of a major British enterprise.

If what you have already heard through the press, radio or TV has made you wonder if the time has come for you to take the plunge into shares this brief guide is for you. Your first reaction may well be surprise – surprise at how easy it is to invest in British Telecom.

How easy?

No harder than shopping by post. You just fill in a simple form – part of the British Telecom prospectus, which will be widely available in November – and send it off with your payment. Or any bank or financial adviser can help you.

How much does it cost?

What you invest is up to you. But the minimum is about £250. Less than half has to be paid out at once. The rest is payable in instalments over about 15 months.

What return can I expect?

You will receive any dividends payable to ordinary shareholders. As a personal shareholder from the start, you will enjoy additional benefits if you keep your shares. You can choose a bonus of free shares, which you will receive after three years. Alternatively, if you are an individual telephone subscriber, you may choose instead to receive (according to the amount you have invested) one or more discount vouchers, worth £18 each, which can be used in part payment of your telephone bills.

Can I check on progress?

The up-to-date price of British Telecom shares will be reported each day in the financial pages of your newspaper. Radio and television also broadcast financial information, so it will be easy to follow any change in value of your shares.

You will appreciate that the price of any share can go down as well as up. But you may be

Figure 4.7 *Magazine advertisement.*

uide to owning com shares.

sure that you will be sharing in the fortunes of a leading company in a vital and growing sector of the world economy.

When do I act?

A prospectus will be issued in November. This is a legal document, carefully describing the activities, prospects and forecast profit and dividend of British Telecom. It will be available at main post offices and banks and widely published in the press.

The prospectus will contain an application form for shares. You must act very quickly.

You never have long to apply. In the case of British Telecom, it could be something like a week to ten days.

So make sure you have made any necessary financial arrangements in good time. If you miss the closing date you miss your chance to get in at the start – and you miss the special privileges of founder shareholders.

How do I find out more?

For more details and your free copy of an introduction to buying and selling shares, send the coupon to P.O. Box 1, Bristol BS99 1BT, or telephone (24 hour service) 0 272 272 272.

Please send me details about British Telecom and share ownership.

Mr/Mrs/Miss ____________________ BLOCK CAPITALS

Address ____________________

____

____

Postcode ____________________

HG1

ISSUED BY KLEINWORT, BENSON LIMITED ON BEHALF OF H.M. GOVERNMENT.

Figure 4.8 *Poster.*

first advertisement in the first issue of a new magazine called *Working Woman*, with the headline 'The Working Woman's Guide to owning British Telecom Shares'. This advertisement drew a huge response for information to the British Telecom Share Information Office. Posters also appeared during the 'Maintenance' phase featuring people who had appeared in earlier commercials (Figure 4.8). Towards the end of the 'Maintenance' period of the campaign, research showed two gaps in public knowledge which were filled by two further television commercials. One explained what a prospectus was and why it was important to apply for one because it contained the application form. The second explained the short application period and urged people to make a decision as soon as the prospectus was available. During the 'Maintenance' phase approximately half a million enquiries for further information and the prospectus were made to the British Telecom Share Information Office.

'Action' phase 5 to 15 November

The 'Maintenance' phase of advertising ended on Sunday 4 November and ushered in a dramatic intensification of the publicity campaign in television, national and regional newspapers and radio in some areas. The 'Action' phase advertisements had four overall objectives—to ensure people knew the entry fee and had their money ready, to instil a sense of urgency by stressing the eight-day window for applications and to continue to merchandise the prospectus for the official application form. It had achieved all this to such good effect that in the two weeks of the campaign over 300 000 enquiries were received.

As can be seen from Figure 4.9, a representative press advertisement, the minimum initial investment of £250 with £100 down was given pride of place. Such a sum was easily within reach of the defined target audience. The eight-day window for applications and freedom from commission and dealing charges for first-time buyers was brought to their attention. As a further benefit, the 'founder-shareholders', those who successfully applied in November would have the choice of either telephone vouchers or a bonus issue after three years over and above the normal dividend payments. The process of application was simple, the form was available in the prospectus to be published on 20 November. For those who had not yet booked their prospectus it was a phone call away—dial 0272 272 272 and book your copy. The usual dish aerial branded the advertisement and it ended with the phrase 'Find out more about sharing British Telecom's future'.

The second press advertisement focused on the twin issues—the time limit (with a prominent headline '8 DAYS') and the prospectus ('Reply to this advertisement TODAY, and we will send you a prospectus as soon as it is published on November 20'). It provided a coupon at the end of the advertise-

Figure 4.9 *Press advertisement.*

ment for a letter response and the telephone number for those wishing to telephone. Simultaneously daily television advertisements marked the countdown to publication of the prospectus and stressed the time constraint on applications.

From 5–15 November, commercial television networks carried a 30-second advertisement, dramatically counting the days until the prospectus day (see Figure 4.10). The scene opened with a calendar specifically entitled 'The British Telecom Share Offer'; a hand flipped over the pages of the calendar to the current date and the voice-over solemnly announced, 'Today is the fifth. British Telecom's prospectus will be published in fifteen days' time on the twentieth'. To hammer home the message, each date of the calendar to the twentieth passed in succession on the screen. The date November 20th was ringed in red and the message enunciated 'It is very important that you receive a copy promptly on publication, because you will then have only eight days in which to come to a decision, make your financial arrangements, and return your application form'. The pages of the calendar then moved on to rest on the key date, 28 November, and the commercial ended with the usual flotation symbol. Night after night, with suitable adjustments of the calendar, the message was repeated.

This series of countdown commercials fulfilled a number of objectives. It clearly focused attention on the passage of time and the key dates, 20 November and 28 November—only eight days in which to come to a decision, make the necessary financial arrangements and return the application form. Equally importantly, it merchandised the prospectus, available even to latecomers by ringing the British Telecom Share Information Office. It did so with good effect. Telephone calls more than doubled from just under 57000 in the week ending 4 November to over 125000 in the first week of the television campaign and to over 150000 in the following week.

'Prospectus' phase 16 to 26 November

From 16 November 60- and 40-second television commercials emphasised the widespread call for share-ownership with the reiteration: 'Parliament made it possible for the people of the United Kingdom to own shares in British Telecom. That means not just big City investors but each one of us'. The advertisements drew attention to the publication of the prospectus on 20 November and its availability to everyone in newspapers, from banks and post offices throughout the country. The broadcasts emphasised that the prospectus carried an application form with a guide on how to fill it in together with the reminder that the closing date for applications was on 28 November. But by this stage, the predicted scale of public response indicated by MORI research was so great that a brake was applied on what threatened to become a massive over-subscription. The golden telephone number 0272 272 272 was omitted.

The
British Telecom
share offer

November
5

November
20

Figure 4.10 *Stills from a TV advertisement.*

		Saturday 17	Sunday 18	Monday 19	Tuesday 20	Wednesday 21	Thursday 22	Friday 23	Saturday 24	Sunday 25	Monday 26
National Dailies	*Daily Mail*	●			●	●		●			●
	Daily Mirror				●	●			●		
	Daily Star				●		●	●			
	Daily Telegraph	●			●	●	●		●		
	Daily Express	●			●	●			●		●
	Financial Times	●			●				●		
	Guardian	●			●			●	●		
	Sun					●					
	The Times	●			●	●	●		●		
	Standard			●	●		●				●
National Sundays	*Mail on Sunday*		●							●	
	News of the World		●							●	
	Observer		●							●	
	Sunday Express		●							●	
	Sunday Mirror		●							●	
	Sunday People		●							●	
	Sunday Telegraph		●							●	
	Sunday Times		●							●	
Daily insertions total		6	8	1	9	6	4	3	6	8	3

Figure 4.11 *Media schedule for the abridged and full prospectuses in the national press. Source: Dewe Rogerson*

Meanwhile, commencing on 20 November the television commercials were joined by a full-scale press advertisement campaign introducing the prospectus at a rate card cost of £3 million owing to the widespread media schedule and the frequency of insertion in the national and regional press (Figures 4.11 and 4.12).

The full prospectus advertisement

On Tuesday 20 November the British Telecom offer for sale prospectus was published across ten pages in four leading national newspapers by Kleinwort Benson Limited on behalf of the Secretary of State for Trade and Industry. There were 3012000000 ordinary shares of 25p face value to be sold at 130p each, with 50p payable on application, 40p payable on 24 June 1985 and the last instalment of 40p due on 9 April 1986. Kleinwort Benson Ltd and Warburg headed the list of 15 underwriters with the other participating banks listed alphabetically.

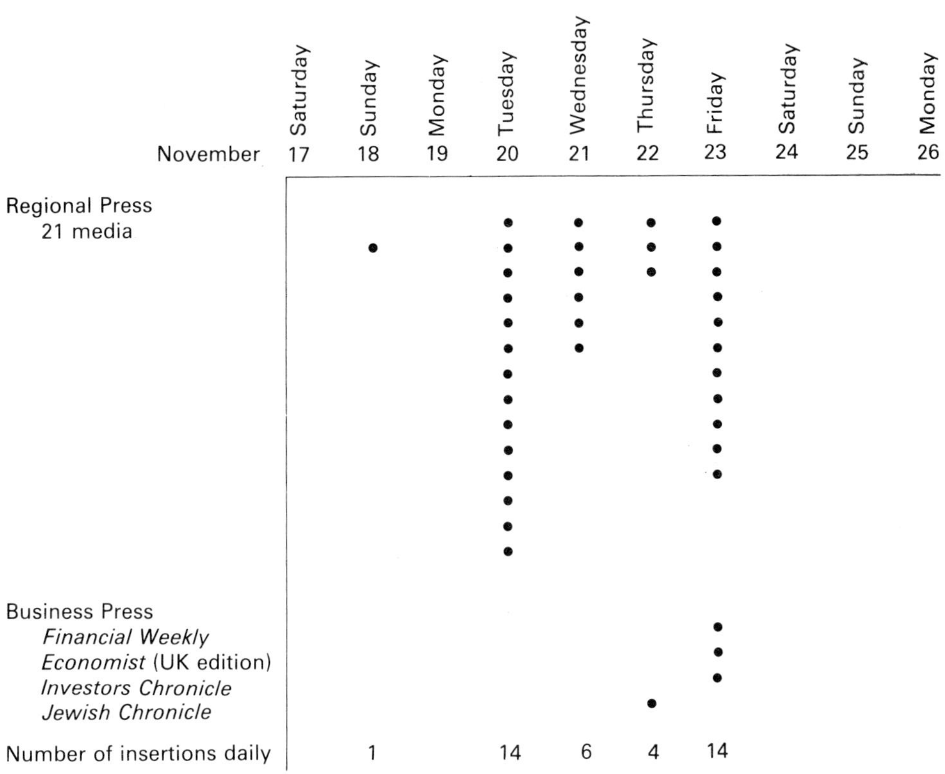

Figure 4.12 *Media schedule for the abridged and full prospectuses in the regional and business press. Source: Dewe Rogerson*

Part One contained the customary introductory information including a definition of British Telecom's business: 'British Telecom is the principal supplier of telecommunications services in the United Kingdom'. The Government as owner of the company was selling 50.2 per cent of the company's issued ordinary share capital. A summary table of historic and forecast profit was followed by details of the actual issue. At the sale price of 130p, the new company's market capitalisation was £7800 million; forecast price/earnings were respectable at 9.35 times and gross dividend yield was estimated at a generous 7.14 per cent. Small investors were guided to Parts Five and Six for a full explanation of the Government's bill-voucher and share-bonus benefits and the special dealing arrangements.

As is customary, the Secretary and Directors of British Telecom were listed together with participating financial advisers, solicitors, and auditors. The registrar bank was Lloyds and the 'charmed circle' of the four brokers to the issue, were Hoare Govett Ltd, Cazenove & Co., de Zoete & Bevan and Scrimgeour, Kemp-Gee & Co.

The second page of the advertisement marks the onset of the small print (figuratively and literally) with all its routine details of the share capital, definitions, admission to Stock Exchange listing, the payment instalment arrangements, the special employee offers and reference to the overseas offerings. There then followed the comprehensive statement of the company's history and prospects required by the Companies Act and the Stock Exchange—details on British Telecom's capital structure, the Government's rights and obligations after the sale as owner of 49.8 per cent of the ordinary shares, and the registration of the Labour opposition's threat to renationalise British Telecom and compensate owners on the basis of the original price of the shares, all accompanied by a detailed assessment of British Telecom covering historical and technological background. This included an account of its business and organisation by division, the constraint on its future pricing policy arising from the terms of its licence and a table of financial results for each activity (where possible). Combined with this were details of British Telecom's regulatory and competitive framework, notably the establishment of competition in telecommunications and the establishment of OFTEL under Professor Bryan Carsberg to monitor and prevent unfair trading. Finally, set out were further details of BT's personnel in terms of numbers and trends in employment, union membership, the employee share schemes and the pension scheme. For the avid reader of the prospectus the past and present position of British Telecom had been laid bare. However, in order to make the investment decision a view of the future and an estimate of prospects was essential, and in section seven of Part One the now familiar appraisal of BT's technological future—so often set out in the preceding publicity— was again spelled out in detail.

Parts Two and Three really belong to the accountants, setting out in extreme detail their analysis of BT's balance sheet, costs, assets, liabilities and profit and loss account. Part Four covered a wealth of information, including financial, organisational, and underwriting details. In Part Five the prospectus returned to further details of the issue, the dealing arrangements, instalment payments, the share benefits and finally the procedure for applying for shares. In anticipation of a huge response, the final page of the advertisement was unusual, devoted as it was to instructions on how to complete and where to send off the application form. The aim was that the instructions should be crystal-clear and the application form user-friendly. The final deadline for applications was firmly stated as 10.00 a.m. on 28 November.

The 'mini' prospectus advertisement

It can be safely assumed that very few people, except professional advisers and very keen core-investors ever read the full prospectus. For the majority of the population the 'mini' prospectus was the key to making their applications. Some 800 000 enquirers to the British Telecom Share Information Office detached the application form from the 'mini' prospectus and returned it with their cheques in the reply-paid envelope. Others relied on the 'mini' prospectus available in Post Offices and bank branches and others used the 'mini' prospectus advertised in the popular and regional press. Far shorter than the full prospectus, at eight pages of A4, the 'mini' gave a brief description of British Telecom and the terms of the offer and included the tailor-made application form with instructions on how to complete and deliver it. The advertisement of the 'mini', covering four pages of the tabloid press, instilled urgency and promised simplicity with the crisp headline 'Only eight days and four pages to decide if you're going to share in British Telecom's future'.

The publication of these full and abridged prospectuses in the press provided information on the product to be sold, the terms of the issue and, equally significant, the vehicle for investment—some 40 million official application forms. The availability and simplicity of the investment had been ensured by skilful design; the provision of information and the distribution of the prospectus had been guaranteed through banks and post offices; the prospectus advertisements had been universally disseminated in national and regional newspapers. But the national availability of the application form was but one half of the equation in an issue of £3.9 billion and a target audience of twenty million people. Crucial to the success of the issue was the return of application forms with cheques by the set deadline. Between the availability of the prospectus and the closing date lay only eight days, itself an extension of the traditional three-day window for receiving application forms. For the sake of a public unfamiliar with such a short time-scale the prospectus phase campaign also contained advertisements in all media to generate *immediate* response.

Whole and half-page advertisements were booked in national newspapers, all aimed to instil a sense of urgency. One advertisement marked the days of the calendar (Figure 4.13); another with the same objective focused full attention on the closing date with the catchy phrase, 'Remember, remember, the twenty-eighth of November' (Figure 4.14). The press advertisements were reinforced by a phased poster campaign announcing the availability of application forms from banks, post offices and newspapers. Commercial radio was also used to alert the public. The radio commercial combined the two sets of messages by announcing that 'Now the offer has been announced. From the twentieth of November a prospectus and application form will be available in banks, post offices and most national newspapers. You will have just eight days in which to apply. If you have decided to share in British Telecom's future the last date of receipt of applications will be the twenty-eighth of November at ten a.m.'

Figure 4.13 *Press advertisement.*

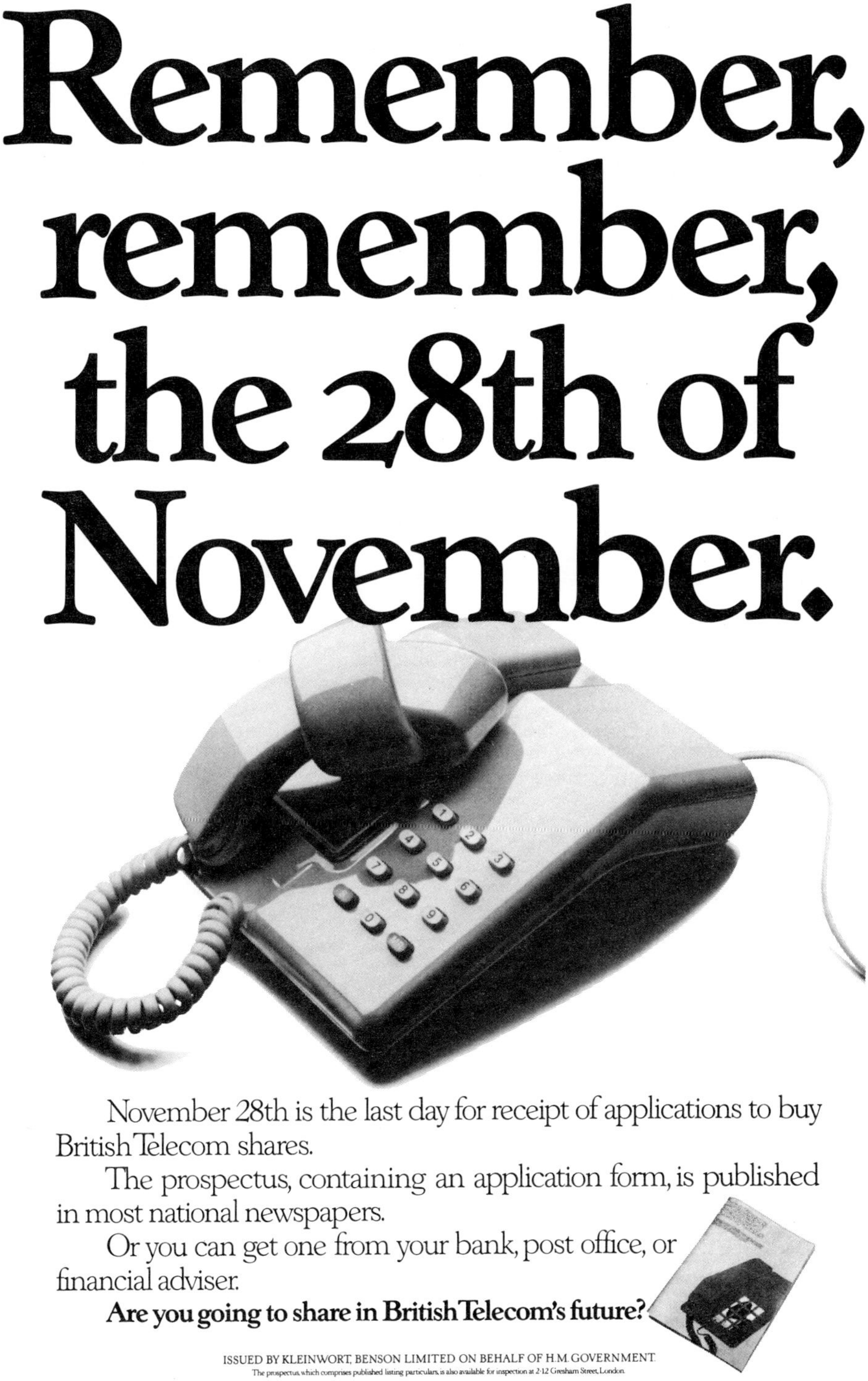

Figure 4.14 *Press advertisement.*

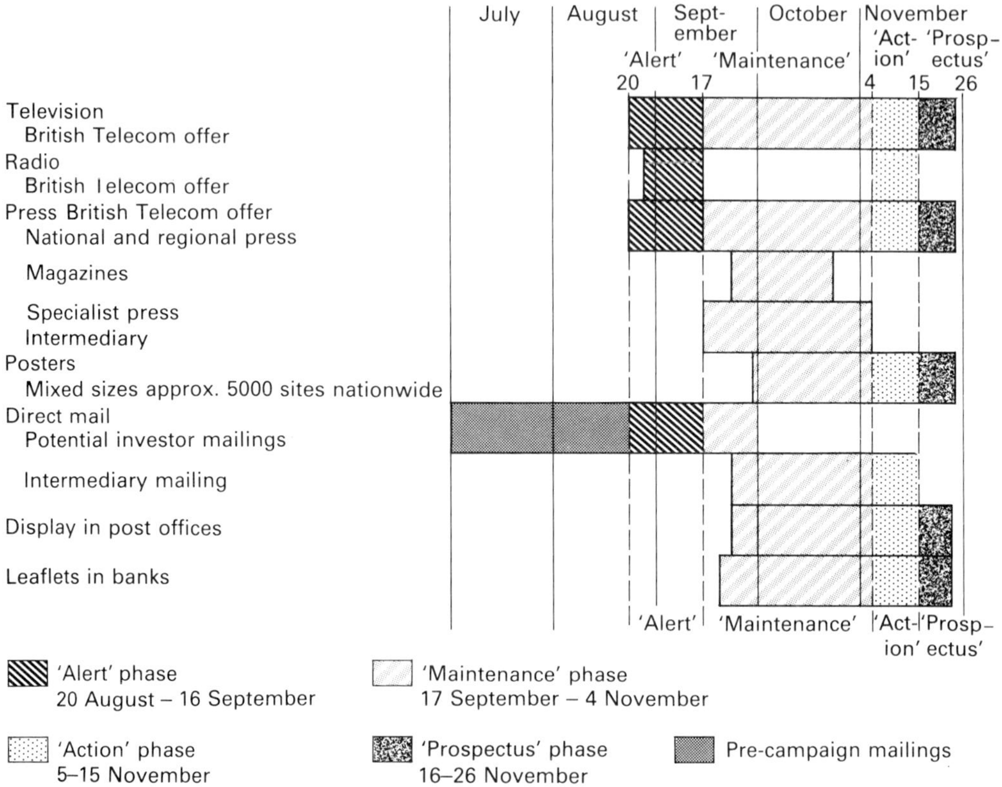

Figure 4.15 *Media mix for the offer campaign.*

MEDIA STRATEGY

The media mix for the offer campaign (Figure 4.15 and Tables 4.3 and 4.4) reflected the desire to create public awareness of the sale of British Telecom, to bring home the realisation that the offer was extended to each and everyone, and to merchandise the prospectus. Television could achieve these objectives most rapidly across the nation, but the press campaign was vital to reinforce the messages, add length to the television campaign, build up a perception of its significance and give weight to the publicity addressed to core investor audiences. The regional press schedule demonstrates the concern to reach out well beyond London, the traditional major market for new issues. The magazine schedule reflects the mass appeal to all socio-economic groups, but above all it represents a clear appeal to the female population, as is evident in its weighting towards female-oriented weekly and monthly magazines. Advertisements in the professional financial press, whilst giving the same messages, were specifically written for this more financially sophisticated segment of the audience.

Table 4.3 *Media schedule excluding prospectus phase*

Popular press	*Sun, News of the World, Daily Express, Sunday Express, Daily Star, Standard, Daily Mail, Mail on Sunday, Daily Mirror, Sunday People, Sunday Mail*
Quality press	*Guardian, Observer, The Times, Sunday Times, Financial Times, Daily Telegraph, Sunday Telegraph* *Jewish Chronicle, Daily Record*
Regional press	*Northern Echo, Liverpool Daily Post, Yorkshire Post, Western Mail, The Journal, Glasgow Herald, Scotsman, Aberdeen Press and Journal, Dundee Courier and Advertiser, Belfast Newsletter, Belfast Telegraph, Irish News, Jersey Evening Post, Guernsey Evening Press and Star, Isle of Man Examiner, Isle of Man Times, Manx Star*
Consumer magazines	(a) Programme publications: — *TV Times* — *Radio Times* (b) Colour supplements: — *Telegraph Sunday Magazine* — *You* — *Sunday Times Magazine* — *Observer Colour Magazine* — *Sunday Express Magazine* (c) Weekly magazines: — *Woman* — *Woman's Own* — *Woman's Weekly* — *Woman's Realm* — *Country Life* (d) Monthly magazines: — *Woman's Journal* — *Woman & Home* — *Family Circle* — *Living* — *Ideal Home* — *House & Garden* — *Illustrated London News* — *She* — *Working Woman* — *Company* — *Expression*
Intermediary Schedule	*Law Society Gazette, Solicitors' Journal, Insurance Age, Banking World, Post Magazine and Insurance Monitor, Stock Exchange Daily Official List, Accountancy Age, Building Society Gazette, Midland Bank Group Newspaper, Barclays News, The Accountant, Accountancy, Financial Weekly, Investors Chronicle, The Economist* (UK edition), *Punch, Management Today, Link-up*
Radio schedule, 4–30 September	
Station:	*Area:*
Capital	London
LBC	London
Chiltern	Bedford/Luton
Hereward	Peterborough
Downtown	Ulster
North West	Lancashire
Scottish	Clyde, Forth, Tay, West Scotland, North, Moray Firth
Tyne-Tees	Tyne and Wear and Teesside
Yorkshire	Yorkshire
West and Wales	Wales and South West

(continued overleaf)

Table 4.3 *Continued*

Midlands	Midlands
Victory	Portsmouth
210	Thames Valley
2CR	Bournemouth
County Sound	Guildford
Suffolk	Ipswich, Bury
Essex	Southend
Piccadilly	Manchester
Manx	Isle of Man

Source: Dorland Advertising

As a result, between 20 August and 12 November the press received a windfall sum of over £2.2 million pounds representing a rate card value of over £3.1 million. The difference in expenditure reflected Dorland's media-buying power and negotiating skills both for obtaining cash discounts and choice position on the page. For the British Telecom Offer campaign (excluding the press prospectus advertisements) Dorland negotiated with media owners for page equivalents within specified dates. This gave the campaign a degree of flexibility with which the size of the advertisement could be balanced against the frequency of appearance.

Posters, an integral part of the media mix, supported the press and television campaign and extended the reach of the campaign literally to street level. The six-week poster campaign, using heavyweight 48-sheet poster sizes on 1200 prime sites a month around major roads, in main terminal stations, airports, underground stations in London and selected provincial cities, added a sense of omnipresence to the urgency of the share offer campaign. The radio schedule also indicated the national scope of the campaign and represented a welcome windfall to the commercial stations. Finally, television commercials at peak-viewing times, including *News at Ten*, *Minder*, and sports and documentary programmes which accounted for over half the

Table 4.4 *Media mix of the share offer campaign 20 August to 26 November 1984*

		%
Press:	Intermediaries	1.25
	Regional	2.75
	Consumer magazines	7.50
	National	21.00
		32.50
Radio		4.50
Posters		9.50
Television		53.50
		100.00

Source: Dorland Advertising

expenditure, ensured a sudden impact and rapid penetration of the target audience on a national scale.

During the campaign, a number of significant changes in attitude were measured by the MORI tracking studies. Stemming from a combination of the intensity, content and execution of the advertising campaign and reinforced by the sheer volume of press coverage, awareness of the flotation of British Telecom reached very high figures—88 per cent of the population. Likewise, BT's image improved substantially on the twin measures of being (a) a good company to own shares in and (b) being a safe, secure investment. Likewise, awareness of the details of the offer increased sharply as the campaign drew to a close. The November date of the offer was known by 83 per cent of the population; awareness of the £250 minimum investment among those interested in buying shares reached 54 per cent, whilst the matching first instalment of £100 was appreciated by 42 per cent of the population. Over half the population recognised the benefits available to private shareholders, as did nearly 70 per cent of those actually interested in buying shares. Awareness of the advertising campaign reached 90 per cent and the golden telephone number of the British Telecom Share Information Office at Bristol was recalled by 16 per cent of the population or over 6.5 million people.

CONCLUSIONS

The privatisation and flotation of British Telecom was accompanied by an advertising campaign of unprecedented scale, marking a radical departure from the traditional low-profile marketing of the typical new issue. British Telecom dominated the advertising sector in 1984. The £16 million 'Power Behind the Button' corporate campaign and the £9 million 'It's-for-YOU-HOO' stimulation campaign assured it of second and tenth place respectively in MEAL's ranking of the most heavily advertised brands of 1984, while the £10.5 million campaign promoting the sale of British Telecom shares by the Department of Trade and Industry lay in eighth place.

Other records were also made. Uniquely, the advertising campaign to launch British Telecom into the private sector was in two distinct parts. The first stage corporate campaign reflected the achievements of British Telecom and from spring 1984, taking account of flotation considerations, included elements designed to increase perceptions of British Telecom as an investment opportunity in which it demonstrably succeeded. The offer campaign, on the other hand, was an unprecedented exercise in selling shares to the nation, and in this too success exceeded all expectations. This campaign was managed by the marketing committee and, excluding the press advertising for the prospectus, was executed by Dorland. This separation of campaigns was essential to comply with the Companies Act and advertising codes: they were separately financed. But although separated by 'ownership' the two campaigns were integrated in purpose. The corporate campaign was the bedrock for shifting image and perceptions as well as attracting interest in British Telecom the company. As Hamish McRae observed in the *Guardian* on 28 June, 'You do not have to be very bright to see that the British market is being massaged by a string of television advertisements leading one to think that Telecom is not a boring old public utility but is actually a high tech, gee-whiz, growth stock'. The subsequent offer campaign followed naturally, to inform, educate and ultimately merchandise the prospectus and application form. The two campaigns were also notable for their length. The corporate campaign lasted for 10 months and was succeeded immediately by the offer campaign of 14 weeks which closed on 26 November. Both campaigns successfully challenged the twin dangers of decay in effectiveness and early peaking. Moreover, the offer campaign itself was the largest direct-response campaign in the world. Every single advertisement invited a response to write in, clip a coupon or telephone for a copy of the prospectus, and ultimately generated a mailing list of 1.3 million. As is customary, every television commercial was submitted to the Independent Broadcasting Authority (IBA) for approval before it was screened. The four phases of the offer campaign broke new ground both for advertising agencies and for the IBA, because it was the first time that the broadcasting media were used to market a new issue. The discussions between the IBA and Dorland were held in strict confidence so it is impossible to guess their nature but for both sides an acceptable solution had to be found. To avoid falling foul of both the IBA Code of Advertising Standards and the Prevention of Fraud (Investments) Act 1958, to forestall cries of share-pushing, each advertisement from the beginning carried a reference to the prospectus, and during the closing stage of the campaign all advertisements clearly stated that the prospectus would be available to everyone, from newspapers, banks and post offices. The slogans used throughout the campaign strike a neat balance between the needs of marketing British Telecom shares and the advertising rules and are carefully phrased with their play upon words: 'Now everyone can *share* in British Telecom's future' to '*sharing* in British Telecom's future' during the 'Action' and 'Prospectus' phases (my italics).

The campaign is also notable for the integrated use of the media mix by which it achieved the multiple objectives of a mass presence, high-profile, depth and ubiquitousness over a sustained period of time. Moreover—unusual for the sale of money products but reflecting the large target audience of some 20 million adults—the campaign moved out of the financial pages to the general pages of the national and regional press and into magazines where it uncharacteristically courted women, with its high exposure in women's magazines.

But perhaps the greatest achievement of the offer campaign lies in its success in advertising a totally unfamiliar product of which the price was unknown practically until the end of the campaign.

Chapter 5

Design

During the BT campaign, Dewe Rogerson placed particular emphasis on design, despite the short life-cycle of the campaign, and appointed two design companies. Liebling Stewart Design Associates were appointed to create a complete mobile roadshow setting including lighting, audio-visual facilities and display panels for the four-week roadshow programme.

Less ephemeral was the appointment of David Pocknell to create a brand imagery for the BT flotation, an image that was distinct but at the same time could incorporate elements of British Telecom's corporate identity. The rationale for this first attempt to produce a separate identity for a new issue belongs to the grand marketing strategy—that is, to create a national event of a financial operation which until now had been in the main restricted to a small segment of the population. All materials associated with the offer for sale were to be clearly and cohesively stamped by one automatically identifiable central image reflecting the essence or core of BT's activities. The result was a deceptively simple line drawing of a telephone—the principal visual symbol, but one in which the buttons provided a versatile series of windows upon which could be superimposed different visual messages to fit each occasion. The device was infinitely adaptable and served many functions as can be seen on the covers of BT's *Products and Services* and the *Power Behind the Button* booklets (Figures 5.2 and 5.3) and in the telephone bill leaflet. All three were distributed in their millions to the general public. The telephone motif was again prominent on the covers of BT's Annual Report and Accounts 1983–84 and the share offer information sheet as well as the 'Mini' prospectus (Figures 5.1 and 5.4). In these instances the Goonhilly dish aerial shows through the button windows and thus provides a direct link to the flotation advertisements. The designer went beyond creating a distinct imagery and established a homogeneity of presentation for all aspects of the literature. Helvetica, a typeface resurrected from its hey-day in the 1960s, combined with standardised layouts in easily digested columns, ensured that all documentation carried the authority of clarity. Throughout the text, the buttons are liberally

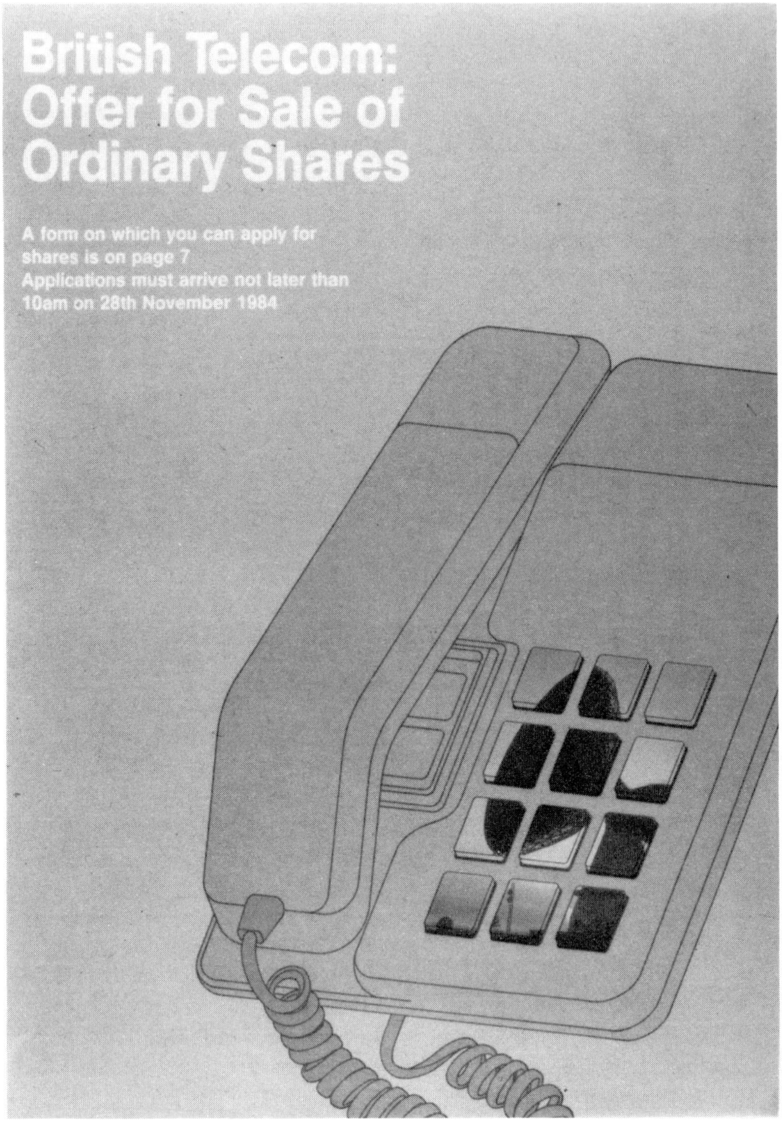

Figure 5.1 *'Mini' prospectus cover.*

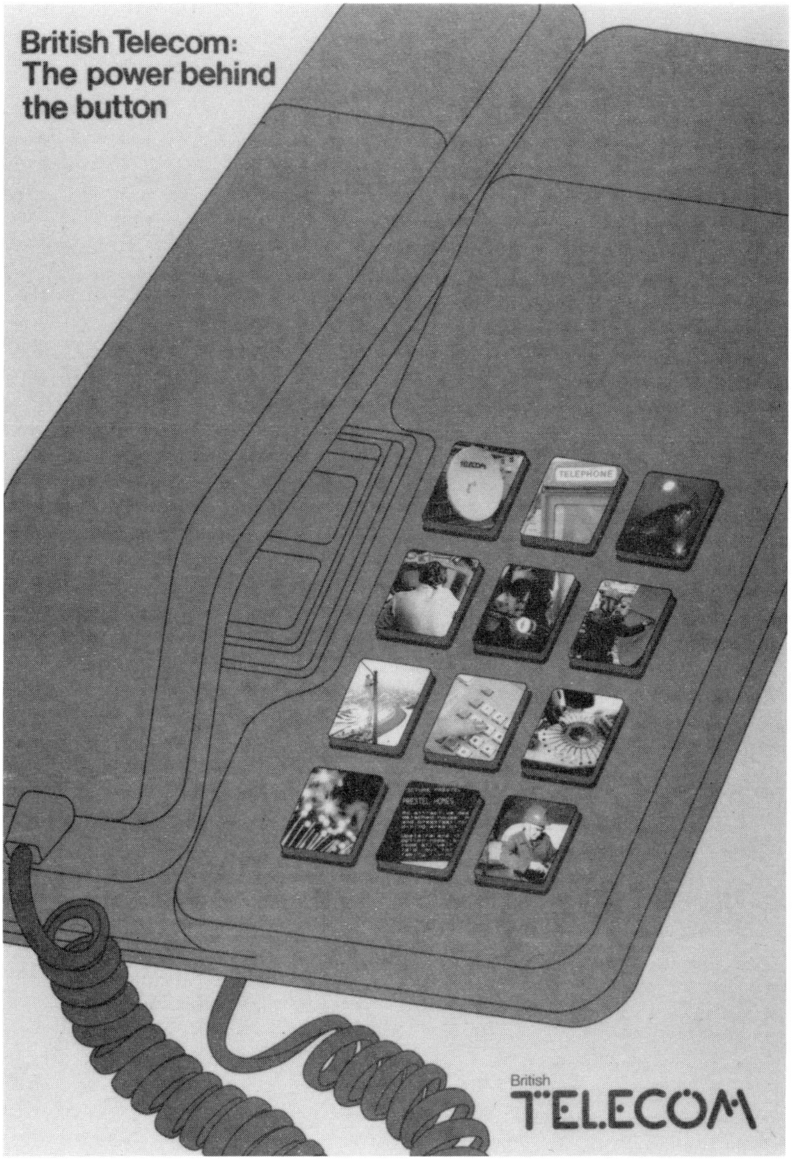

Figure 5.2 *'Power Behind the Button' booklet cover.*

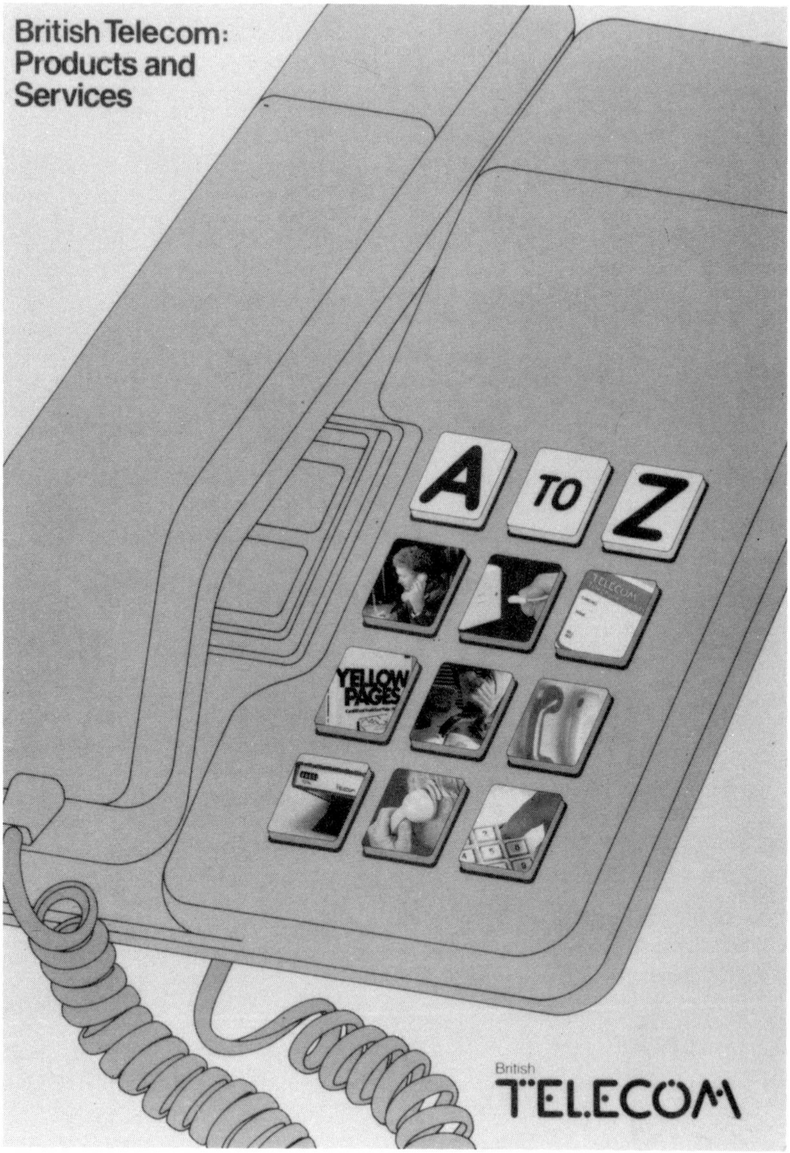

Figure 5.3 *'Products and Services' booklet cover.*

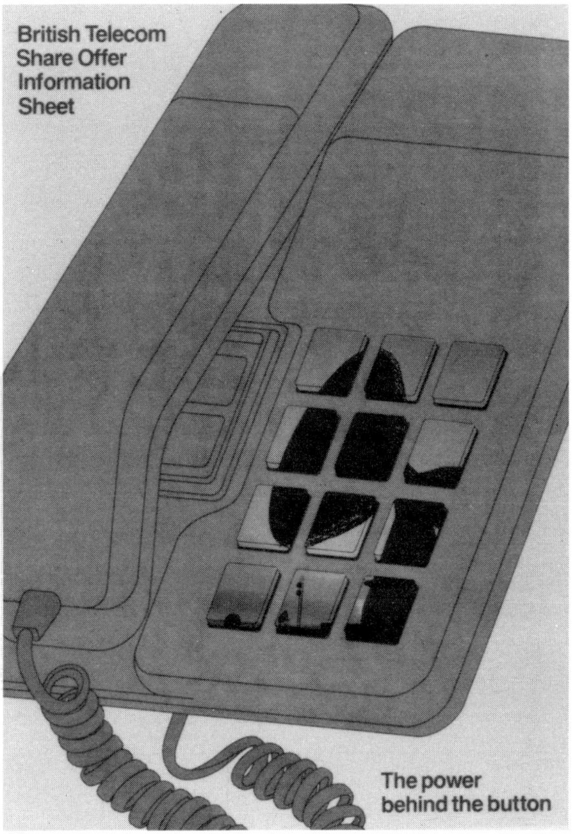

Figure 5.4 *Share information sheet cover.*

applied to act as visual message carriers, incorporating either illustrations, numerals, or simply acting as highlighters of paragraph beginnings. However, it is in the prospectus that the art of design really came into its own. The application form, with the seven steps which investors had to take so clearly marked by numbered buttons, marks a radical departure from traditional application forms (Figure 5.5). Further inside the prospectus and mainly for connoisseurs are the tables and text. The figures are presented in clear, unambiguous style designed for legibility and ease of use. The impact of the layout owes much to the determination to present the requisite information with style, a process that took eight or nine design draft copies but resulted in a model of simplicity and approachability.

The adopted brand symbol of the telephone and buttons echo the 'Power Behind the Button' theme of British Telecom's corporate advertising campaign and thus skilfully linked the corporate identity with the flotation. That was the essential strategy of the marketing committee, and to that extent design secured their purpose with distinction.

British Telecommunications plc
Application Form

To:
The Secretary of State for Trade and Industry
Kleinwort, Benson Limited
Lloyds Bank Plc
British Telecommunications plc

I/We attach a payment for the amount payable now of £ ______ **1**

and I/We offer to purchase ______ **2**

Ordinary Shares in British Telecommunications plc at 130p per share on the terms and subject to the conditions set out in the full prospectus, comprising the listing particulars, dated 16th November 1984.

For the bill vouchers put "YES" in Box A. For the share bonus put "YES" in Box B. COMPLETE ONE BOX ONLY.

A. Bill Vouchers ☐ **OR** **B. Share Bonus** ☐ **3**

Signature ______ **4**

USE BLOCK CAPITALS **5**

Mr., Mrs., Miss or title | Forename(s) (in full)
Surname
Address (in full)
Postcode

→ ☐ **Pin here your cheque/bankers' draft for amount in Box 1**

Fill in this section only when there is more than one applicant **6**

USE BLOCK CAPITALS

Mr., Mrs., Miss or title / Forename(s)	Mr., Mrs., Miss or title / Forename(s)	Mr., Mrs., Miss or title / Forename(s)
Surname	Surname	Surname
Address	Address	Address
Postcode	Postcode	Postcode

7

Signature	Signature	Signature

FOR OFFICIAL USE ONLY
1 Acceptance No
2 Shares accepted
3 Amount received £
4 Amount payable £
5 Amount returned £
6 Cheque Number

Figure 5.5 *Application form.*

Chapter 6

Logistics of Printing and Distribution

Given the size of the target audience and the nature of the message it was essential that the twin channels of information dissemination and distribution be widespread. MORI research had indicated that 44 per cent of those expressing an interest in British Telecom shares regarded the post office as the most convenient purchase point. Whilst this was thought inconceivable to put into practice the 1560 Crown post offices in the UK could, nevertheless, substantially reinforce the marketing campaign and bring it to the attention of their ten million weekly customers. At a fairly modest cost the Crown post offices would extend the marketing drive as a channel of information and as a means of merchandising the address and telephone number of the British Telecom Share Information Office in Bristol. For this purpose from 24 September all post offices would display posters, and from 15 October all would carry leaflet dispensers for the public fact sheets, and in 500 of the largest post offices with a video service BT's 40-second television commercials could be transmitted. Coinciding with the start of the 'Prospectus' phase on 20 November, the leaflet dispensers would be filled with the 'mini' prospectus, delivered in bulk from the printers directly to each post office.

DISTRIBUTION OF LITERATURE THROUGH THE BANKS

David Clementi, of Kleinwort Benson, acting on behalf of the Government in the sale of British Telecom shares, envisaged the banks playing an integral part in the retail marketing campaign. The banks were a familiar and established presence in the High Street with over 14400 branches. Research had shown that 50 per cent of the general public see their bank manager as the principal source of financial advice and information. Moreover, given the scale of the issue and potential applications it was vital to enlist the support of the banks not only for the distribution of information but for answering the public's questions and for the distribution and receipt of application forms. However, on the question of giving financial advice or taking a view on the British Telecom issue the banks were reluctant to change their custom and practice and therefore refused to recommend an individual share.

From 6 August Kleinwort Benson began negotiations with the Head Offices of the London Clearing banks—Barclays, Lloyds, Midland, National Westminster, Williams and Glyn's, plus the Yorkshire Bank, Royal Bank of Scotland, Clydesdale and TSB at general manager level. All in all, twelve banks were approached. The banks agreed to distribute information, answer public queries, make available the full prospectus and the 'mini' and take in completed application forms in return for a commission of 1.5 per cent on successful applications submitted directly bearing only the banks' stamp and of $1\frac{5}{8}$ per cent for applications submitted through a local stockbroker carrying two stamps, that of the bank and the broker.

In order to answer the public's questions, the bank at branch level had to be made aware of the background of British Telecom and be conversant with the details of the offer and how to apply for shares. Thus bank branch managers were included in the direct-mail campaign to financial intermediaries and received invitations to attend the PLC train seminars and roadshows. Regional stockbroker co-ordinators included branch managers in their marketing drive and as a matter of course the banks were the recipients of all the public documentation.

In September 1984 all branch managers received five copies of the British Telecom introductory memorandum for banks together with 20 fact sheets on the share offer. The memorandum for banks was for internal use only and included an application form for additional information packs. In total, 24 000 information packs were sent to bank branches from the British Telecom Share Information Office in Gresham Street, London, and thus nearly three-quarters of a million fact sheets were available to the public through this medium. In addition, to prepare bank tellers, copies of a video featuring TV star Noel Edmonds and covering the likely questions that the public would ask were made available to each bank by the last week of September. In November banks were to play a vital role in the distribution of prospectuses and in taking in completed application forms.

LOGISTICS FOR THE FULL AND 'MINI' PROSPECTUSES

The announcement of the British Telecom share offer price on Friday morning, 16 November, initiated the final stages of the largest and most complex financial printing and distribution process ever undertaken. Over the weekend, Burrup Mathieson and Co. Ltd co-ordinated the printing, mailing and delivery of some ten million items of print, consuming more than 350 tons of paper, to be distributed from a variety of outlets to millions of people.

The printing, mailing and delivery of the 'mini' and full prospectuses was the culmination of a ten-month project on behalf of HM Government, British Telecom and Kleinwort Benson, involving the production of millions of copies of some 80 different documents in support of the biggest sale of shares in world history. Over the weekend of 16–18 November 400000 copies of the full 64-page prospectus and over six million copies of the eight-page 'mini' prospectus were printed and distributed to key collection points. The 15 print works and 8 mailing houses were supervised by 50 senior Burrup managers to ensure that quality, consistency specifications and production deadlines were being met. The 'mini' prospectus was mailed on Monday by first-class post to arrive on Tuesday, 20 November on the doorstep of all those who had sought information from the British Telecom Share Information Office. Around 6000 bank branches and all Crown and sub post offices throughout the UK received deliveries of both types of prospectus and approximately 320000 British Telecom employees and pensioners each received a copy of the 'mini' prospectus and personalised share application form. According to Peter Barker, Chairman of the Burrup Printing Group, 'This project has been more of a management and logistic project than a pure printing job. We have had to exploit the full planning, setting, proofing, printing and distribution experience and resources of the six companies in the group to ensure we meet the demanding technical specifications and timetable required. The commitment to absolute confidentiality and security which characterises so much of our work has also been a vital ingredient.'

Part of the Extel Group, 356-year-old Burrup Mathieson are the largest and one of the most highly respected financial printing companies in Europe. They were selected by Kleinwort Benson—the contract was won partly on price and partly on the basis of Burrup's expertise not only in commercial new-issue documentation but on its work for several privatisations including British Aerospace, Cable and Wireless, Britoil and Jaguar. The quotation for typesetting and printing the prospectus was based on the expectation of 25 stages of proof-revision, equivalent to the full resetting of the prospectus a dozen times and in the event this proved an underestimate. Ancillary documents such as newspaper advertisements, Extel cards, letters of acceptance, regret and rejection were included in the quotation, each with its own specification and expected levels of proof-revision.

As the chairman indicated, the project extended beyond the usual mechanics of typesetting and printing numerous documents to fundamental logistical problems of timing, co-ordination and distribution. Planning was an essential ingredient and occurred at two levels. Internal company conferences, held both before and after regular discussions with Kleinwort Benson, began with monthly meetings and ended with weekly meetings to thrash out last-minute problems. Planning the logistics of distribution involved sophisticated computer programmes to ensure individual labelling, numbering and addressing.

The task was stupendous given the quantities involved. For example, the preliminary distribution of the prospectus involved 30 000 parcels delivered to more than 20 000 different locations.

Even a routine function such as the ordering of paper became a complex operation due to the nature of the offer for sale. For many of the documents quantities were directly related to the level of demand, and whilst this could be forecast close to a deadline there were no means of assessing the quantities of paper required in Spring 1984, at which time the paper order had to be placed. For example, Burrup began with a print-run of one million acceptance letters but had to more than double it at the last minute. After the conception in June 1984 of the 'mini' prospectus, the early projection of an 80-page prospectus and 700 000 copies was changed to 400 000 copies of a 64-page full prospectus. For strategic reasons, paper had to be bought in advance to ensure sufficient quantities and to contain costs. Careful planning and monitoring allowed Burrup to react quickly to the changing requirements. Paper amounts to approximately 50 per cent of the cost of production and paper pulp is priced in US dollars, a currency which showed an upward spiral in its value against the pound sterling in Spring 1984. The quantity required was quite staggering. The prospectus alone required the equivalent of one reel, 7400 km long and 450 mm wide, the pulp equivalent of 1200 trees! The choice of paper supplier was not easy. In the end Burrup appointed Bowater, one of two major UK paper suppliers.

The printing and distribution of the two forms of prospectus were in themselves huge tasks, and complicated by the need for speed. Given the huge quantities involved and the requirement for national distribution it was realised, early on, that Burrup alone could not fulfil the print run. Outside printers would have to be found and parts of the operation sub-contracted. The process of approaching printers, assessing their capabilities, negotiating terms and appointing them was no easy task, for Burrup found themselves in a sellers' market, dependent upon printers with regular magazine print runs anxious not to disturb their regular schedules. Figure 6.1 is taken from the planning stages of Burrup's operation and, although different from the final outcome, shows the type of geographic spread of 'mini' prospectus printers, full prospectus printers and mailing houses each with their own catchment area who were appointed to ensure speedy national distribution.

Both prospectuses were type-set by Burrup Mathieson, a process accompanied by the traditional proof-reading, resetting overnight and delivery at 8 a.m. the following morning for yet another proof-reading by the customary army of lawyers, bankers, brokers, accountants, public relations men, totalling some 60–70 people. After numerous proof-readings, Burrup developed master films of each prospectus and arranged for printers to collect two copies of the film under strict security from its headquarters, Crane House.

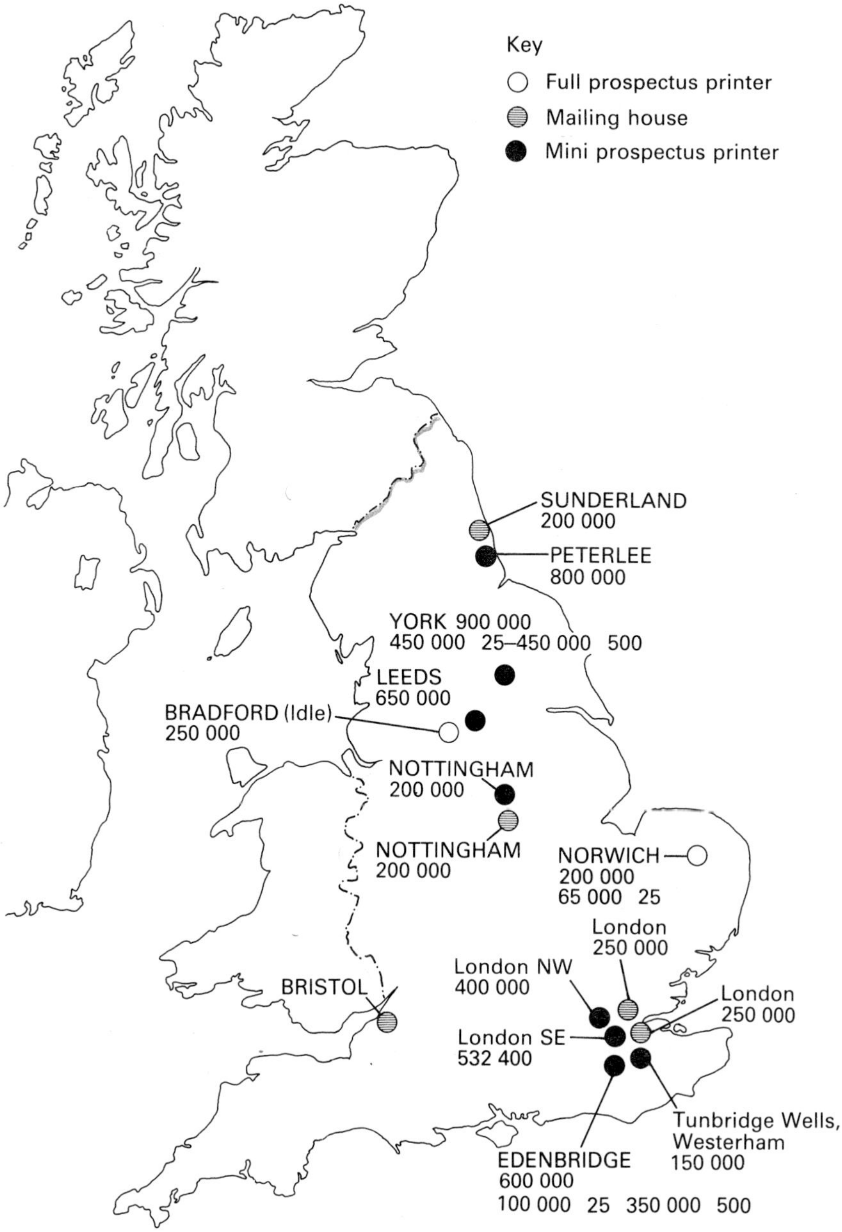

Figure 6.1 *Printing and distribution. Source: Burrup Mathieson.*

Two strategic sets of documents, requiring different delivery cycles posed additional problems and the varied quantities required by bank branches added another difficult dimension to the logistics of distribution. Crowning the enquiry-handling programme based on the British Telecom Share Information Office in Bristol, was the 'mini' prospectus, sent with a covering letter and reply-paid pre-addressed envelope to over 1.3 million homes. Printed in regional centres the 'minis' were despatched to a neighbouring mailing house where they were inserted into envelopes specially prepared from computer listings to reflect the six alphabetical groupings of enquiries, and delivered to the post office for overnight delivery.

There was in addition the complicated task of getting the 'mini' to the banks, post offices and financial intermediaries. It was made easier by computers matching quantities to codes for each branch and by internal deliveries made by the banks and Post Office themselves. Less complicated was the provision of the 'mini' to around 80 000 BT pensioners and 240 000 employees accompanied by their special application forms.

The scale of the tasks of planning, organisation, co-ordination, printing, collating and delivering such quantities and numerous items to their appropriate destinations can be seen in Tables 6.1 and 6.2. The quantity of 'mini'

Table 6.1 *Distribution*

Principal distribution centres	Full prospectus quantity	'Mini' prospectus quantity
1. Receiving banks	189 550	1 614 700
2. Other banks	16 180	458 700
3. Post offices (collected and distributed by post office)		
(a) Crown post office (1600)	41 500	800 000
(b) Sub post offices	21 000	500 000
4. Mailing		
(a) Employees and pensioners	—	310 000
(b) Mailing list	—	1 100 000
(c) Bulk delivery to Bristol	—	300 000
(d) Bulk delivery for mailing to financial intermediaries	—	177 500
5. Telecom offices/factories (100)	14 000	24 000
Sub total	282 230	5 284 900
6. Underwriters and brokers in the City (14)	42 000	14 000
Kleinwort Benson	10 000	100 000
7. Brokers to the offer	20 000	450 000
8. Regional co-ordinators	4 250	622 000
9. Other brokers and special delivery	125	380 100
10. Contingency	35 395	113 500
Total quantity	394 000	6 964 500

Source: Kleinwort Benson

Table 6.2

Full and mini prospectus			Employee and pensioner documents	Other documents
Wednesday 14 November	8 a.m.	Distribute final proofs of full and mini prospectuses		Distribute proofs of newspaper advertisements and Extel card
	6 p.m.	Proof to printers of full and mini prospectuses		Proof to printers of newspaper advertisements and Extel card
	8 p.m.	Main proof-reading session at Crane House full and mini prospectuses		
	10 p.m.	Approve all pages without price information on full and mini prospectuses and (a) make films for main runs (b) Order underwriting proofs without prices		Approve all parts not affected by price. Order final proofs
Thursday 15 November	8 a.m.	Underwriting proofs without prices of full prospectus available	a.m. Main proof-reading session	a.m. Main proof-reading session
	6 p.m.	FIX PRICE Proceed with all pages of full prospectus without prices Proof to printers of: (a) full prospectus — all pages with prices (b) mini prospectus	FIX PRICE Proof to printers of: (a) Red and blue application form (b) Guide to red and blue form (c) Covering letter to red and blue form (d) Guide to red form (e) Covering letter to red form (f) Pensioner's covering letter (g) Pensioner's application form	FIX PRICE Proof to printers of: (a) Newspaper advertisements (b) Letters of acceptance (c) Extel card

continued overleaf

Table 6.2 *Continued*

Full and mini prospectus			Employee and pensioner documents	Other documents	
Thursday 15 November continued	9 p.m.	READING PARTY	READING PARTY		READING PARTY
	11 p.m.	Approve all pages of mini prospectus Approve all pages with prices of full prospectus and: (a) Make films (b) Order underwriting proofs with prices (c) Order press copies	Approve the above seven documents		Approve newspaper advertisements
	Midnight	Reading parties in Norwich and Bradford (to check first 32 pages of the full prospectus).			
Friday 16 November		IMPACT DAY			
	2 a.m. to 4 a.m.	Full and mini prospectus printers collect film from Crane House		8 a.m.	Circulate final proofs of Extel card and letters of acceptance
	6 a.m. to Noon	Reading parties at printers to check running sheets (a) Full prospectus Norwich and Bradford (b) Mini prospectus — Peterlee, York, Leeds, Pershore, Nottingham, Edenbridge and two in London		10 a.m.	Approve Extel card and letter of acceptance Deliver plates and bromides to newspapers. Block pulls to Kleinwort Benson and Dewe Rogerson and lawyers
	8 a.m.	Distribute underwriting proofs with prices of full prospectus			
	Late p.m.	First bound copies of full prospectus available			
	6 p.m.	Commence distribution to mailing houses, Securicor and the Post Office			

Saturday 17 November	8 a.m.	Continue distribution	8 a.m.	Affix labels to pensioners' application forms Mailing house commence inserting employees' and pensioners' documents		
	p.m.	Despatch final copies of mini prospectus to mailing houses				
Sunday 18 November	6 a.m.	Full prospectus — complete despatch to the Post Office				
	Noon	Mini prospectus — complete despatch to the Post Office				
	p.m.	Complete printing and binding of both full and mini and despatch final copies to Securicor				
Monday 19 November	3 p.m.	Post Office collect final envelopes from mailing houses	Noon	Despatch employee documents to BT offices/factories	Noon	Approval required to release Extel card into the Extel service
Tuesday 20 November		ADVERTISING DAY 5 million 'mini' prospectuses: (a) 1.3 million through letter boxes of share information office enquirers (b) 4 million available from banks, brokers, post offices, regional co-ordinators, (c) British Telecom mailing (d) 381 000 full prospectuses for merchant banks, stockbrokers, receiving banks, other banks, post offices, regional co-ordinators, British Telecom				Newspaper advertisements 40 million application forms appear

Source: Burrup Mathieson

prospectuses was huge but the allocation of these documents was the result of much consideration and was designed to ensure both that legitimate demand would be met in full while still keeping the costs down as far as possible. The achievement of this printing and distribution task would have been difficult under any circumstances, but especially so in the case of British Telecom where the whole operation took less than 84 hours. It was made possible by Burrup's in-depth planning, management and monitoring throughout the operation resulting in all documentation being printed and in position within the required timescales as well as the magnificent efforts alike by the printers and the Post Office whose postmen delivered with 95 per cent accuracy over 1.3 million envelopes through Britain's letter boxes on Tuesday morning, 20 November.

Chapter 7

Public Relations

During the marketing programme, below-the-line promotion was to become just as vigorous and prominent as the above-the-line advertising campaigns. Public relations was the essential ingredient to the creation of the British Telecom flotation as a national event. It was deployed behind the 'Power Behind the Button' campaign, to sustain the image of British Telecom as a very large company poised on the threshold of a technological revolution. At the same time it was vital to the share offer, its terms and timing being accompanied by an intensive press relations campaign. These two objectives—the promotion of British Telecom the product and the offer terms—were interdependent and each campaign supported and sustained the other through their shared audiences, techniques and central control by the marketing committee.

THE PRODUCT—BRITISH TELECOM

Following the announcement of privatisation British Telecom had to begin building bridges with new audiences in order to establish a high profile. Coinciding with and underpinning the corporate advertising campaign, lay British Telecom's own major public relations programme. Major highlights of public relations activity included a series of exhibitions, competitions and other items designed to educate target segments to British Telecom's new corporate image.

Building on the world telecommunications exhibition in Geneva in October 1983, to which BT flew, at its own expense, selected journalists, MPs, major customers and staff, came the Guildhall Exhibition in late April 1984 which projected the new British Telecom to City customers, financial journalists and institutional investors and was used as an occasion to brief bankers, stockbrokers and investment analysts about the company. In September, British Telecom mounted an Open Week at Martlesham, its research and development laboratories in Suffolk. It was an exciting combination of 'Telecom today', featuring products and services, and 'Telecom tomorrow', which presented the scale and breadth of the company's research and thereby reinforced the research and development section of the forthcoming prospectus. Demonstrations showed how optical fibres could be installed into ducts in offices without disruption to office routines. Displays included coloured optical fibre cables being blown through clear hollow tubes running for more than a hundred yards around a laboratory. Cable television with a video screen showed all the services available to cable-television customers. Special trains hired by British Telecom took a total of 3000 people to see this exhibition, including MPs, industrialists and business people, educationalists and civil servants. The exhibition had just two days in which to show the City institutions and the media BT's research depth and sophistication. Designed to inform and transform cynicism about BT's efficiency and management Martlesham was a success and as a sales drive these three exhibitions alone are estimated to have increased sales by £12 million.

In addition there were more general national public relations campaigns including the much publicised competition for the 'golden voice', the winner becoming the voice for British Telecom's speaking clock service. The regional finals took place from late July to mid-October, generating much local press publicity, and the announcement of the winner in December created additional national publicity for the company. In June British Telecom mounted a picture exhibition of 'The People Behind the Button', but more far-reaching was Telecom Focus, a nationwide exercise with the theme 'caring for the customer', ranging from demonstrations of hardware such as phones, callmakers, switchboards and telex machines to the public opening of telephone exchanges and stands at town fairs, carnivals, exhibitions and agricultural shows. It is estimated that more than five million people visited British Telecom premises on open days and countless more visited their stands. For the schools, British Telecom ran a nationwide competition inviting young people to devise new methods of using their technology to help the disabled and elderly, supporting teachers with a library of videos, books and briefing materials. The production of the 20 millionth telephone was turned into an event and publicised by a *Daily Mail* competition, and to extend the British Telecom presence further the company issued a stream of news releases on their activities. Finally an event gaining worldwide media coverage, neatly placed a fortnight before publication of the prospectus, was the inauguration by HM the Queen of ANZCAN, the major new submarine cable. The events described above are a representative sample of British Telecom's public relations activity, focusing mainly on its corporate image for a variety of audiences.

FINANCIAL PUBLIC RELATIONS FOR THE OFFER FOR SALE

The other area of intense public relations activity centred upon the marketing campaign and the share offer itself. Two broad lines of activity predominated.

Firstly, there was an active financial communication programme directed at City institutions, fund managers and investment analysts conducted by the British Telecom board, Kleinwort Benson and Hoare Govett. Secondly there was a sustained and intensive press relations campaign punctuated with large-scale press conferences undertaken by Dewe Rogerson and the British Telecom Share Information Office (London) on behalf of Kleinwort Benson to concentrate attention and editorial coverage upon specific events and news announcements. It was supplemented in the closing weeks of the offer campaign by daily telephone conversations with City editors.

Education and the establishment of new channels of communication with new audiences were the hall-marks of financial public relations activity conducted by and on behalf of British Telecom. To overcome the dearth of knowledge about BT's commercial scope and organisation and to advance financial advisers along the learning curve well ahead of the issue, a novel technique was employed to provide an in-depth, well-researched background to the industry and the company. Thus in February began BT's presentations to three of the newly appointed lead stockbrokers on all aspects of their business. The result was three independent stockbroker reports. The first report from de Zoete & Bevan was published on 31 May and was followed by a full-scale press conference the next day to obtain publicity for its comprehensive analysis of BT's business. In July reports by Hoare Govett and Scrimgeour, Kemp-Gee became available for mass distribution within the City and among financial intermediaries and opinion formers.

Similarly, the publication of British Telecom's annual report and accounts for 1983–84 marked the first major step in the process of the City actually scanning the financial status and accounting practices of this public giant. Distribution of the document and its accompanying twelve-page news release was widespread. The attendant press conference was for the first time heavily weighted towards the financial press in order to explain the slight downturn in profits and provide a firm grounding on the achievements and progress of British Telecom.

In order to reinforce the information flow to City audiences, British Telecom ran an intensive and sustained programme of meetings with stockbrokers, investment analysts, and fund managers. For example, the British Telecom board during the marketing drive gave 110 presentations to financial institutions. From May, every Wednesday and Thursday, began a series of lunches for major City institutions and fund managers at BT's headquarters. In addition to such fixed and formal events the board kept in regular telephone contact with the City. For example, Douglas Perryman, Director of Finance, estimated that he had lunched or spoken to 1500 City analysts, investment managers and stockbrokers during the flotation campaign! In essence such programmes not only gave British Telecom an opportunity to educate the City in all the intricacies of its business well before the publication of the prospectus, thereby building up familiarity with the organisation in the City, but created a continuous dialogue with an essential but hitherto unknown audience and provided a platform for BT's management team to project its potentialities.

PRESS RELATIONS

In summer 1984 the British Telecom Share Information Office, located in Room 426 in BT's headquarters in Gresham Street and equipped with direct telephone lines to Kleinwort Benson, Hoare Govett and the Department of Trade and Industry, opened its doors for business. The office was the visible, physical focus of the public relations media offer campaign acting on behalf of Kleinwort Benson and staffed by Telecom employees. It was managed by John Harben of Dewe Rogerson and Peter Wynne Davies of British Telecom. The functions of the office were to control the information flow and distribute legally cleared information to the previously identified 1300 journalists. The objective was to educate and interest all sections of the media through regular briefing reports, news releases, dissemination of specific news stories and prepared articles. It was designed to obtain editorial comment and features in order to build awareness and interest among the public and, of equal importance, to generate a large volume of third party endorsement.

The first major task of the BTSIO (London) was to open the eyes of journalists to the reader/viewer appeal of the new issue. Subsequently the task became one of sustaining that interest and building up a peak of editorial and news flow to coincide with the 'Action' and 'Prospectus' phases of the marketing drive. MORI research provided quantifiable evidence of widespread public interest in the share sale, demonstrating as early as July that around 36 per cent of the population was interested in British Telecom shares. Moreover, as financial journalists knew from previous experience, their editorials and advice were highly valued by the public when making a saving or investment decision. The extent of public interest, when combined with journalistic self-interest, proved a sufficient lever to initiate media interest. In order to sustain this press interest the BTSIO (London) issued a string of news releases on marketing tactics and news items relating to the issue. In addition, journalists were automatically supplied with news stories tailored to their audience and every three weeks the BTSIO (London) report provided research findings and a brief commentary on published offer facts, as and when they became available after announcement in the House of Commons. To ensure the provision of information was, however, not enough. It was necessary to try to shape the outcome and to this end the BTSIO (London) organised a series of media briefings ranging from personal to group sessions. The effect of such news management was notable for the extent of coverage and perhaps the best example of press relations can be seen in the four feature articles which appeared in the *Financial Times* in October, tailor-made for the City. They thoroughly reviewed BT's new management outlook, the state of the business and range of its activities, its industrial relations and the details of the offer in the context of UK equity issues. But to capture the financial journalist and business journalist was not enough. It was essential to spread media coverage of the issue from the City pages and business programmes into the more generalised areas of family finance pages, documentaries and news programmes. To this end, specially tailored briefings for each medium and audience were undertaken. For example, on 9 October, Dewe Rogerson organised a Family Finance media briefing which was attended by journalists from most of the national press, Channel 4 Business Programmes, LBC, and the financial media of the small saver such as the *Investors Chronicle, Money Magazine, Financial Weekly* and *The Economist*. It was hosted by Dewe Rogerson, with speakers David Clementi on the issue details and Anthony Carlisle on the research, marketing and communications programmes. MORI research findings helped to highlight the extent of public interest. The intention was not only to extend coverage of the story but to emphasise the short period of time available for applications so that journalists could alert the holders of 7-day and 28-day building society accounts to organise their withdrawal of funds. The briefing was followed by a formal question and answer session, chaired by Roddy Dewe, and concluded with informal briefings over lunch.

Another function of the BTSIO (London) was the channelling of questions to the right people. Thus matters of Government policy were automatically directed to the Department of Trade and Industry, those relating to the company, to British Telecom and queries on the overall marketing campaign and preparation for the flotation to Kleinwort Benson and Dewe Rogerson.

The central provision of information ensured control over the volume, content and timing of the information flow and formed the foundation of news reporting, editorial features and commentary for an audience of millions of private individuals, thousands of financial intermediaries and hundreds of fund managers, stockbrokers and investment analysts.

PRESS CONFERENCES

The successful press conference is the result of meticulous attention to planning, logistics, timing and detail. These ingredients, when combined with real news and a professional presentation, can make or break a press conference and no new issue was in greater need of a successful press conference launch than this one.

But for the very recent precedent of the Jaguar flotation barely three months before, the British Telecom Pathfinder prospectus for the offer for sale of 3012 million ordinary shares would have been unique in the British new issue market, though quite normal practice in America. The Pathfinder contained all the information and terms normally published in a full prospectus other than the price at which the shares are to be offered, though in this case the reference to a 7 per cent yield gave a clue to the final price and resulted in accurate editorial guesses of the price of the shares some three weeks in advance of its publication.

Thus, this press conference certainly had some hard news to present and the launch of the Pathfinder prospectus on 26 October 1984 certainly rewarded the organisers and resulted in a breakthrough in the television news and feature coverage of British Telecom and its forthcoming debut on the stock market. From this day the flotation became a national news story on the front and general pages of the press, in the news on radio and television. The launch of the Pathfinder was a model of successful news management and ensured a rising crescendo of third-party discussion and endorsement in the vital four weeks leading up to the closing date of the offer.

The location of the press conference in the Grand Hall at the Connaught Rooms was ideal in three ways. It was easily within reach of Fleet Street and the City, it was a prestigious venue and it was blessed by a large high-ceilinged dramatic room. This sense of theatre was reinforced by the seating arrangements organised in a semi-circle facing an empty raised stage. All visual material, including press packs, badges, room-dressing and set-design were in harmony, branded by the flotation symbol. The Dewe Rogerson press packs contained the Pathfinder prospectus, a covering press release and a note to editors (see Appendix 7.1) giving story leads. The presenters were Martin Jacomb, Lord Rockley and David Clementi representing Kleinwort Benson, and Sir George Jefferson and Deryk Vander Weyer, chairman and deputy chairman respectively of British Telecom. All had undergone a full rehearsal the previous day. Of the two hundred guests for the launch of the Pathfinder, half were journalists. They can in fact be divided into three distinct groups. Firstly, the specialist City-business journalists, whose main interest would lie in the technical details of the company such as the balance sheet, pension fundings, inter-divisional financing and the profit and loss account. Secondly, the generalist City-business journalists, including family finance journalists, news reporters, diarists, feature editors and television reporters whose interests would lie less in technical details and more on aspects of the offer dear to potential investors. This group of general news journalists would be decisive if the story of the British Telecom launch was to reach the public. Finally, from the trade and technical magazines, Telecom-watcher journalists from *Computer News*, *Electronics Weekly* and *Marketing*; each would have their own interest, but in common with the generalists their stories would reach the public with big bold headlines and selling by-lines. The remaining guests were made up of representatives from various departments in BT. In addition there was a diverse group of advisers, and specialists from S. G. Warburg, Dewe Rogerson, Hoare Govett, Cazenove, the Department of Trade and Industry, the Treasury and Morgan Stanley, lead managers for the American sale of British Telecom shares.

At 11.15 a.m. the press conference began with the arrival of the British Telecom/Kleinwort Benson team on the platform and after a two minute photocall Martin Jacomb introduced the team. Sir George Jefferson gave a short presentation tracing the history of the company since its separation from the Post Office and summarised parts of the prospectus covering general information about the company and its capital, constitution and relationship with the Government. He outlined BT's activities and sketched the regulatory and competitive framework and moved on to specific financial information such as the company's financial position and profits. One phrase in his speech denotes a marked change in philosophy: 'The interests of shareholders and customers are not in real conflict. In order to make sure that shareholders prosper we must satisfy our customers, by making them feel we offer an attractive service, good value at the price'. BT was announcing in effect the shift in emphasis towards becoming a market-led company to match its transformation into a private sector company. To David Clementi fell the task of briefing the media on the marketing of the offer and the mechanics of the flotation. He told his audience that three key dates lay ahead, namely Friday 16 November (the day when the price of the shares would be announced), Tuesday 20 November (the day on which prospectuses would be advertised and distributed) and Wednesday 28 November (the day the offer closed, thereby leaving only eight days for the general public to apply for shares). He announced that this day's conference marked the start of a major roadshow programme to last three weeks, touring both the UK and overseas financial centres. Martin Jacomb then opened the session to questions from the floor and the panel were joined by advisers. At 12.30 the session ended but interviews were held with selected journalists. Thereafter the principals and their advisers moved on to BT headquarters and were available to telephone enquirers fed from the BTSIO (London).

Just as crucial as the launch of the Pathfinder prospectus was news management of 'Impact Day', 16 November, when Kleinwort Benson announced the share price. To gain the maximum publicity for the two news stories, namely the announcement of the British Telecom share offer price and the underwriting of the offer, a mix of strategies was employed.

For the announcement of the share price there was a rather unusual gesture, the unfurling of a vivid yellow rooftop banner on Kleinwort Benson's Fenchurch Street building, a scene covered by all UK television networks from an adjacent roof-top of Barclays Bank. To satisfy the demand for pictures there was the customary photocall session, featuring Sir George Jefferson, Deryk Vander Weyer, David Clementi, Martin Jacomb and Lord Rockley, the key architects of the new issue poised on Barclays roof against the backdrop of the banner announcing the price.

For details of the price, announcement of the underwriting terms and placing of the issue there was a press conference, held in Kleinwort Benson's boardroom and attended by 40 journalists hand picked by Dewe Rogerson to ensure widespread publicity in all major media. With Martin Jacomb in the chair, David Clementi and Lord Rockley made short statements to the select audience and there then followed a short question-and-answer session, ending with television and radio interviews. In addition, there was the essential press statement drafted by Roddy Dewe and hand-delivered to all City and news editors including London-based correspondents of the regional media, the Press Association and Reuters Financial for international coverage, Extel and TOPIC for the City institutions, giving all the necessary details.

Given the intrinsic importance of the news and the careful organisation of media it is not surprising that there was extensive coverage. For example the Saturday coverage in the family finance pages pointed out the advantages of the offer based on the unanimous view that the price was attractive and editorial coverage bordered on the ecstatic with headlines predicting a massive oversubscription. In the event, such expectations were proved correct. When the share application lists closed at 10.01 a.m. on 28 November the Government seemed to have achieved its long desired wider share-ownership objective.

Apart from the major press relations campaign undertaken by the BTSIO (London) and Dewe Rogerson, Kleinwort Benson as the issuing house hosted lunches with the press in the run up to the publication of the Pathfinder prospectus, designed to move opinion forward from an understanding of British Telecom to appreciation of the offer for sale itself. In discussions with financial journalists Kleinwort directors gave indications of the level of the yield and other technical elements, especially stressing the argument that BT shares would be highly competitive with a building society account. It was the financial journalists' editorial coverage of this information which was absolutely vital in converting interest in British Telecom to purchase.

RESULT OF PRESS RELATIONS

The result of the combined BT, Dewe Rogerson, Kleinwort Benson and Hoare Govett press relations activity can be measured crudely by the volume—a 12-inch pile of press cuttings and several hours of air-time on radio and television. But the measure of a campaign lies not only in the quantity, but more significantly in the quality of information and its impact. The press relations campaign visibly changed media perceptions on the outcome of the share flotation and achieved widespread third-party endorsement of the offer terms.

In June and July, coverage of British Telecom was essentially confined to the City pages of the national press and focused on the terms of BT's licence and the outcome of the flotation against a background of falling stock markets in which the FT Industrial Ordinary Index fell from its peak of 922 points in May to a nadir of around 750 points in July. The tenor of editorials was largely critical of the central tenets of the marketing objectives. In one issue *The Economist* questioned whether American investors would take to British Telecom shares 'of what is, after all, a utility company' (9–15 June 1984). Margaret White of the *Guardian* posed the question whether the huge flotation of 51 per cent of BT at the end of the year would succeed (11 June 1984) and Jonathan Davies of *The Times* echoed similar uncertainty, for '. . . in the next nine months alone the Government is committed to selling shares in Jaguar, British Telecom and British Airways. It had been hoping to raise up to £10 000 million from its asset sales programme over the next four years. These proceeds will have to be downgraded . . . unless stock markets show signs of recovery' (11 June 1984) whilst David Meyrick, writing in the first issue of *Broadcast*, noted the dissonance between the corporate advertising campaign, demonstrating microchip exchanges and nano-seconds, and the reality of crackling telephones, dilatory exchanges and fully engaged major trunk routes. And casting a further cloud on the strategy of a simultaneous issue of BT shares in America there was the current example of the Reuters issue being made simultaneously in London and New York where lack of US demand at the time of the issue had adverse effects on the pricing. Real doubts on the success of the issue owing to its sheer scale were expressed by David Malcolm, chief investment manager of the Royal Insurance Company (29 June, Channel 4). In July with the London Stock Exchange's indices still sliding, the *Daily Telegraph* observed, 'The City is getting distinctly edgy about the sheer quantity of stock the Government is preparing to push its way' (5 July). Whilst the City's lack of enthusiasm was known to the organisers it must have been a disappointment to them to hear M. Jancowski, investment analyst of Simon Coates, talking on BBC's *Financial World Tonight*:

> 'I would be very surprised if private individuals would be interested in taking up shares. I think the problem really rests on British Telecom's reputation and many people will look at their latest phone bills and the increase that they've seen over the past few years and feel more or less reluctant to really put an investment into British Telecom. The other problem as well is that alternative investments offer very good rates of return such as Building Society deposits, and given that British Telecom is risky, why not go for safety?' (19 July)

The *Daily Star* encapsulated pervasive doubts in the headline 'Flop fears for the £8 bn super sale' (23 July).

But in August press journalists began to throw off their negative reservations and focused on MORI research findings responding sympathetically to the advertising campaign and sensing the growing scale of public interest thanks to the corporate campaign. On 21 August the *Standard*, London's evening paper, marked the turn around with the assertion that the issue would be favourably priced, investors would be unlikely to lose money and concluding, 'Since almost everyone can recognise a good thing when they see it, the issue should succeed in attracting a wide spread of ownership'. From September onwards favourable press comment was extensive, with headlines such as 'Telecom's buy call has a winning ring' in the *Sunday Telegraph* (9 September). A crucial moment came with John Bell's article headed 'Telecom heads for markets on a super 17 per cent yield', terms which far outweighed returns on alternative investment options (*Sunday Express*, 23 September). On 10 October, *The Times* reported to its expanding audience, attracted by the 'Portfolio' game, that 1·2 million adults were seriously thinking of buying £250 worth of BT shares, and towards the end of the month commentators vied with each other in forecasting the likely uptake of the shares, whilst immediately prior to the publication of the prospectus, the *Sunday Times*, *Daily Star*, *Daily Mail* and *Daily Express* all forecast an oversubscription. These sentiments of enthusiasm eclipsed the lone voices of caution, ranging from Mr. Walker of the Bank of England who expressed some scepticism about the likelihood of an early revival of the personal investor and Edgar Palamountain, long-time chairman of the Wider Share Ownership Council, who claimed the Government's objective would not be achieved and observed, 'The use of public money to promote share-ownership in general is one thing; its use to bribe people to take up a particular issue is quite another'. (*Financial Times*, 11 October). Nevertheless headlines remained almost unanimous in their enthusiasm:

'British Telecom share offer too good to miss'	*Daily Telegraph*, 26 September
'Telecom set for £3500 M share rush'	*Daily Mirror*, 27 October
'A million new capitalists? How to get in on the act'	*Sunday Times*, 28 October
'It's the dream sale of the century'	*Daily Express*, 30 October
'Buzby share-out is the one for yoo-hoo'	*Sun*, 30 October

Journalists also vied with each other to explain the attraction to small shareholders and advise purchasers of the share sale as a 'once in a lifetime chance for small investors to make money'. Security, a good return, an opportunity to make money, all these factors were used to recommend the shares to the public, whilst MORI figures for the mounting enquiries at the BTSIO (Bristol) reassured writers as to the outcome and in a sense fuelled their enthusiasm. The avalanche of enthusiastic media comment and advice was reinforced by the *Mail on Sunday*'s competition with British Telecom shares as the prize, whilst the *Sun*'s page-three girl appeared three times with the telephone as an accessory. High and low, BT was everywhere.

But what of the institutional investors who were expected to take up the bulk of the offer? Fund managers, investment analysts and stockbrokers were not immune to the ever-increasing crescendo of enthusiasm in the media and, as the Harris/*Observer* Poll discovered, 80 per cent of fund managers polled stated that they intended to buy British Telecom shares in early November. The *Standard* noted the convergence of public enthusiasm and media coverage in its impact on the City with the observation, 'The City loves a winner'.

The predictions came true. 2.3 million successful applications were recorded, scaling down of applications was necessary and the Government had achieved its objectives. The shares opened at a premium of 90 per cent on the partly-paid price and 33 per cent on the full offer price and the commentators' verdict on the marketing programme was almost ecstatic:

> 'British Telecom was a monumental corporate finance exercise and, in almost every respect, a monumental achievement'
>
> Lex, *Financial Times*, 4 December

> 'Never has there been such a City launch. Never has there been such a stampede to buy shares'
>
> *Daily Mail*, 29 November

> 'The biggest, best publicised and most highly stage-managed share sale in history was put to the test in the marketplace yesterday and received with rapturous approval'.
>
> *Daily Telegraph*, 4 December

Appendix 7.1

The launch of the Pathfinder — Dewe Rogerson's Pathfinder press release

NOTES TO EDITORS

BT Pathfinder Prospectus

1. **Planned timetable**

Friday 16th November	—	Announcement of Offer price.
Tuesday 20th November	—	Prospectus and mini prospectus published.
Wednesday 28th November	—	10 a.m. Offer closes.

2. **Financial Information**
   Key information on profits, dividends and offer statistics are set out on page 1.
3. **MORI research findings September/October 1984**
   80% of UK adult population aware of privatisation.
   25% (11 million people) interested in BT share purchase. This of course, does *not* mean that all these people will buy shares. It does indicate, however, the wide pool from which the BT private investor will come, particularly when compared with the small existing private shareowner base — 1.8 million people, of which 600,000 own shares in only one company (most likely in employee schemes).
   The interest in British Telecom shares is broadly spread amongst all sectors of the public and all regions of the country.
   Although to begin with, men were significantly more interested than women, their interest is now virtually equal. It looks as though, for many, buying British Telecom shares will be a joint decision.
   Clearly, the great majority of people thinking of buying British Telecom shares have never bought shares before. Consequently, they want a good deal of information about what they have to do, how and when. They also want to know what it's going to cost them and what returns they can expect.
   Over 90% of those interested want information and intend to turn to banks, professional advisers; Post Offices (main branches of which have fact sheet on the offer) or the Share Information Office.

The information the public wants includes: (MORI)

Likely profit to be made on the shares	(44%)
How long the shares have to be held	(39%)
The cost	(34%)
How to buy shares	(33%)
Future prospects of British Telecom	(33%)
Where to get information	(32%)

It is clear that information about how to go about buying the shares is of equal interest to potential investors as questions relating to the terms of the offer and the returns that may be expected.

4. Incentives

Choice of telephone bill vouchers or free shares offered by Government *only available to individuals who apply for the initial offer.* Details are on the attached information sheet.

5. Minimum Investment & Staged Payments

The minimum investment will be roughly £250 (depending on the precise price of the offer) and the payments will be staged over 16½ months approximately as follows:

November 1984	—	40%
24 June 1985	—	30%
9 April 1986	—	30%

Individual and institutional shareholders will be eligible to receive the full dividends paid by BT during a period when the shares remain partly paid.

6. Availability of prospectus and mini prospectus

Prospectus and mini prospectus, together with application forms, will be widely advertised and available from banks, post offices, stockbrokers and other professional advisers throughout the country. The mini prospectus will also be posted to all those who have contacted the British Telecom Share Information Office (800,000+ as at 25/10/84).

7. Short 8 Day Period During which Offer will be open

The duration of an offer for sale is necessarily short. Those interested in applying for shares will therefore have to move quickly and, if using the post, send the cheque and application form by first class post by 26th November at the latest, so as to arrive at the receiving bank by 10 a.m. on Wednesday 28th November.

8. Application Form

Instructions on how to complete and deliver the application form have been made as simple as possible and a specimen is included at the end of the Pathfinder.

9. Special Selling Arrangements

While the terms of the incentives provided by the government for the individual founder shareholders are clearly designed to encourage people to hold onto their shares for a period, special arrangements have been made to reduce normal buying and selling costs for small investors to 31st December 1986.

10. BT Employee Offers

Most of BT's 240,000 employees will be eligible to take up a three part special offer from the government. First approximately £70 worth of shares is being offered free. Next, for an investment of up to £100 the employee will receive twice this amount in free shares. Finally, employees can buy additional shares, up to a value of £2,000 at a 10% discount.

Well over 800,000 people have so far telephoned or written to the British Telecom Share Information Office.

11. Enquiries to Share Information Office

Over 850,000 people have so far telephoned or written to the British Telecom Share Information Office for information.

12. Par Value

The prospectus mentions that 3,012 million ordinary shares of '*25p each*' are to be sold. This is the par or nominal value of a share and is not to be confused, of course, with the price at which a share will be offered for sale, which will only be announced when the prospectus is published.

Q.

"The Government intends to offer shares in British Telecom for sale later this year. How interested do you think you will be in purchasing such shares?"

		Profile of those interested in British Telecom shares (October 1984—Source MORI) All interested = 100% Percentage of population: 25%	*Profile of existing shareholders in this country.* (Source: NOP Financial Research Survey) Percentage of population: 5%
1.1	*Class*		
	AB	20	56
	C1	26	26
	C2	32	11
	DE	21	6
1.2	*Sex*		
	Male	49	57
	Female	51	43
1.3	*Age*		
	18–24	17	5
	25–34	19	13
	35–44	20	16
	45–54	15	21

	55–64	19	19
	65+	11	27
1.4	*TV Region*		
	Scotland	7	11
	North East	4	3
	Lancashire	13	11
	Yorkshire	10	6
	Midlands	18	10
	Wales/West	4	7
	East Anglia	10	7
	London	26	26
	Southern	9	14
	South West	1	5
1.5	*Telephone in household*		
	Yes	89	Not available
	No	11	
	Data is from a representative quota sample of adults (18+) in 154 sampling adults. Interviewing took place between 14–17 October, 1984. In total 2080 people were interviewed, of which 525 said they were interested in BT shares.		Data is from the NOP Financial Research Survey based on 18 national probability samples interviewed between April–September 1983. In total 33 307 people were interviewed.

Source: Dewe Rogerson.

Chapter 8

BT Employee Share-Ownership

> 'Employers who are also owners will correctly perceive the absolute identity of interest between themselves and the success of the companies for which they work'.
>
> Mr. John Moore, Financial Secretary to the Treasury, to Wider Share Ownership Council Forum, 3 October 1984

The concept of profit-sharing through employee share-ownership schemes is an importation from America and is based upon the assumption that employees who share in the profits generated by their company also share a common interest or purpose with capital, namely the success of the business. In the UK the concept received the Labour Government's backing with the introduction of tax relief for employee profit-sharing schemes which the Conservative Government, elected in 1979, later extended. By June 1985, 900 employee schemes existed involving 750000 employees and their share options are valued at over £1 billion.

The growth of such schemes owes much to legislation and the programme of privatisation. Legislation has created three different types of schemes. The first introduced by the Labour Government in the Finance Act of 1978 created employee share schemes available to all employees. Subsequent Finance Acts made the initial scheme more attractive by increasing the annual limit on the value of shares which could be allocated and at the same time reduced the compulsory period for which the shares would have to be held if they were to escape payment of income tax. In 1982 the Conservative Government introduced savings and related share schemes and two years later the executive or 'key' employee share option scheme was unfolded. By increasing the types of schemes and their attractions, employee share-ownership has expanded rapidly. Similarly, the programme of privatisation itself added to the momentum. The privatisation of 12 state-owned companies added 330000 employees to the ranks of the employee share-owners, but British Telecom set the record for employee participation. Ten per cent, or 301200000, British Telecom shares were reserved by the Government on favourable terms for eligible company employees and pensioners. Three different offers were available. Firstly, the free offer, was universally available on application for a free gift of 54 shares; secondly, for employees who invested their own money there was a two-for-one matching offer which for an investment of £100.10p bought 77 shares and carried 154 free shares making a total of 231 shares or £370 worth of shares for around £100. A combination of these two packages of free and matching offers was internally known as the blue offer, mirroring the colour of the application form. The third, or red, offer gave each employee the right to apply for up to 1600 shares at a discount of 10 per cent on the price at which the shares were to be offered to the public and applications would be met in full. £100 million-worth of shares were reserved for employees who applied on the relevant red application form. These shares, just like those bought by the public, carried a choice of telephone vouchers and bonus shares. The discount offer was a complete innovation made possible by a change in tax law introduced by the Finance Act 1984.

Given such a generous offer-package to employees it is not surprising that 96 per cent of the eligible staff ignored their unions' opposition and applied for shares. Over 222 000 of the 238 000 staff applied for the free offer of 54 shares and some 84 per cent took advantage of the matching offer. However, a much smaller proportion of employees ventured into the red form priority application. A little over a quarter of British Telecom's employees entered for the priority/discount shares available on the red form, and a little under a third of British Telecom's pensioners applied for shares on their special priority pink form. Nevertheless, the generosity of the offer ensured that British Telecom had a substantial shareholding amongst its employees.

THE BT EMPLOYEE SHARE COMMUNICATION PROGRAMME

Employing a mix of channels, including letters from the chairman, information booklets, articles in the house magazine, *Telecom Today*, and works presentations, the hall-mark of the campaign was the provision of information so that employees would understand the benefits of each offer and how and when to apply. All employee communications were branded by the jigsaw logo, specially designed to bring instant recognition that the information concerned the share offer. The logo has been retained to distinguish future share information to employees. Educational and informational in content the communication programme only once came close to offering any advice, and then it was by an independent journalist writing in the June issue of *Telecom Today*; he suggested that the free offer came 'as an undisguised gift horse', and it was, therefore, 'foolish not to accept as you have absolutely nothing to lose' and as for the matching offer: 'if you can find £100 you're prepared to lock up . . . I would do so.'

As early as May 1984, when the details of the employee offer were announced in the House of Commons, employees received a letter from the company's chairman, giving the conditions of eligibility and details of the employee share offer and ending with the observation, 'Finally I would ask you to give very careful thought to these opportunities to acquire shares in your company. You should make sure that you read all the information which will be made available to you over the coming months, so that your decisions are based on the fullest possible knowledge of the proposals'. In effect the board welcomed the shareholder arrangements and was concerned that employees were given sufficient information upon which to base a decision. Each employee received the chairman's letter, drafted, printed and enveloped in advance and posted immediately after the announcement in Parliament signalled by radio-paging from a van outside the House of Commons. In July all employees received, like the public, the booklet, 'The Power Behind the Button'. In September employees received the booklet on employee shares and another covering letter from the chairman. Under an attractive, glossy cover, with the jigsaw logo, the guide described the employee share offer and explained the important features of the employee share-ownership scheme. Each section was colour coded—blue for the free and matching offers, red for the priority/discount offer and yellow for general information. The guide offered definitions of unfamiliar terms, explanations of unfamiliar concepts such as 'what is a share?', 'share prices', 'buying and selling shares', 'dividends' and the rights of shareholders, plus the major special features of the offer, including instalment payments and incentives. The red offer was again set out in detail and for the blue offer, or the employee shares, employees were assured that their applications, unlike the public's, should be met in full. The trust arrangements and income tax position were spelt out in full with simple examples of typical situations such as selling the shares after two years, and selling after four years, both at an enhanced value. Finally, in the sleeve of the jacket, employees were given advance notice of the flotation pack, due to be sent to their homes in November, and the urgency of completing the application forms and allowing at least two working days for delivery before the closing date was stressed.

In addition to the written material, the Employee Communication Unit was responsible for the organisation and administration of presentations throughout the organisation—no easy task for a company with almost 240 000 staff scattered throughout the UK. To overcome the logistical problems, 350 employee shareholding liaison officers were appointed in July in each of British Telecom's major divisions and trained by the Employee Communication Unit. From September, these officers ran seminars, working from standard briefing materials and a video provided by the unit. The presentations, available to all employees in work time, consisted of a short video entitled 'Are you getting your shares?' to explain the offers. This was followed by individual practice sessions on how to fill in the application form for the blue and red offers using dummy application forms. These forms were as close to the real application forms as possible given their early preparation over two months before the flotation. This exercise, designed to familiarise employees with the compulsory form-filling at one level, was also designed to concentrate the mind on the decision of whether to invest and how much to invest so that when the time came they had only to fill in the form. The practical hands-on part was then followed by a question-and-answer session chaired by the liaison officer, using the question-and-answer pack prepared by headquarters, and for convenience divided into sections covering general questions, those relating to the offer, application procedure, payment terms, the trust and trustees, and selling and transfer of shares and dividends. It is noteworthy that presenters were warned not to stray from the official prepared answers or to attempt to answer a question on which they had not been briefed. Such instructions ensured that presentations were standardised in format, content and range. The seminars all ended on an administrative note to facilitate the construction of the BT employee share register. Quite surprisingly, in view of its corporate advertising focusing attention on information technology and computerisation, British Telecom itself lacked a central register of employees! Instead each local area maintained its own lists, predominantly manually recorded on paper. In late 1983, in preparation for privatisation, British Telecom had instituted a £6 million programme to transfer paper records on staff to a new computer system, the personnel-related information system for management, or PRISM. For the employee share scheme, Hill Samuel Registrars Ltd. needed a complete list of staff, their names and addresses. Transfer of such employee data to Hill Samuel had begun in May 1984 in order to build up a central register for direct mailing. To ensure that all details were correct and up-to-date, staff attending the seminars were requested to check their personal entries. This exercise gave British Telecom headquarters the first central register of all its employees.

In November, to keep up the momentum of employee communications, employees received a further letter from the chairman and the accompanying Employee Shares Special Bulletin bringing an up-to-date version of the share offers and introducing information on the BT Employee Sharesave Scheme or 'green' offer to be made separately after the flotation of British Telecom in November. It announced the board's decision to introduce a save-as-you-earn share option scheme and set out details of staff eligibility needed to acquire further shares in the company on favourable terms.

Finally, on Monday 19 November, Hill Samuel on behalf of British Telecom despatched to all eligible employees' homes the flotation pack, consisting of the 'mini' prospectus, the personalised blue and red application forms with respective guides and a reply-paid first-class envelope. The pack was despatched at the same time that the 'mini' prospectus was sent to over a million members of the public. Simultaneously, distinctive posters appeared on company notice-boards in order to publicise the despatch of the application forms and instruct staff on what to do if they did not receive their forms.

The employee communications programme reflected the fundamental policy decision to separate the employee offer programme from the retail marketing programme targeted at the general public. Conducted by British Telecom, for British Telecom, it appeared detached from the Government-sponsored policy of public sale. This strategy arose partly from the desire to remain aloof from the controversial political debate and partly from the decision to consolidate company–employee relations.

The programme was marked by a substantial media mix of written, audio-visual and personal presentations producing a regular flow of information, as and when it became available, to all employees either at work or at their homes. In common with the retail marketing programme the campaign was centrally planned and co-ordinated using a standardised approach and subject to the routine problems of design, logistics, drafting, vetting and printing processes requiring enormous attention to detail and frequent consultation with lawyers and the Department of Trade and Industry. However, whilst the communication programme deserves applause on its own merits, the employee-response probably owes more to the generosity of the offer terms.

Chapter 9

UK Institutional and Overseas Markets

Kleinwort Benson Limited was the co-ordinator and manager of the offer for sale of 3012 million British Telecom shares on behalf of HM Government and had the usual task of any issuing house in planning strategy and co-ordinating a multidisciplinary team of specialists in order to achieve a full subscription to a simultaneous UK and overseas offering of shares. A team of eight, headed by Lord Rockley, based on a functional division of labour with David Clementi in daily control, devoted themselves to the mechanics, terms and marketing strategy of the new issue aided by Warburg as BT's financial advisers. In this momentous task they were aided by the appointment in January 1984 of four stockbroking firms to the issue sometimes known as the 'charmed (or inner) circle'. Hoare Govett Ltd. was appointed as the lead broker to the issue on behalf of the vendors, HM Government; de Zoete & Bevan, and Scrimgeour, Kemp-Gee & Co. for their strong position in the technology and telecommunications sector of the market, and Cazenove & Co. were appointed to represent British Telecom. Though it is normal to have only one lead broker to an issue, a precedent of two for the largest issues to date had been set, and given the scale of the task of selling £3.9 billion-worth of British Telecom shares and the need to penetrate deeply into the institutional investor market the need arose to support the inner circle with an outer circle of another dozen stockbrokers to prime the market, and to talk to the institutional investors and their advisers well ahead of the issue and in the months leading up to it. Closer to the date of the flotation the inner circle would advise the issuing house on the price of the shares based on their discussions with the institutions and on the day of the issue the inner and outer circle of stockbrokers would offer a predetermined amount of shares to their institutional and private clients. Together, the inner and outer circle of brokers covered some 80 per cent of the volume of the market and were thus well placed to distribute the shares amongst the investment community of over 300.

There were three elements in the institutional campaign. Firstly, there had to be discussions with the financial institutions which started immediately in late January, beginning with one-to-one talks with some three dozen investing institutions and extending to some 300 by the time of the flotation. In the early months of the year the 'charmed circle' began a series of general discussions around the size of the issue and the manner of the offer. There was some preliminary testing of underwriting capability and assessment of the potential investment interest. The feedback from such preliminary meetings was not encouraging. Many of the institutions were cautious or luke-warm and some were completely hostile so that there were serious doubts about the possibility of doing the issue at all! Whilst an issue of £3.9 billion represented only some 25 per cent of the institutions' cash flow it nevertheless exceeded the £2.5 billion invested in UK equities by the institutions during 1983 and would have to compete with other investment opportunities such as overseas investment. In the event, Government statistics show that overseas investment by financial institutions was lower in 1984 than in 1983. Moreover, in its June 1985 quarterly bulletin the Bank of England observed that in the second half of 1984 capital issues by companies dropped sharply, probably because firms were unwilling to tap the market in the run-up to the BT flotation.

In March the results of MORI surveys among institutional investors, specialist technology stockbrokers and financial journalists further deepened the gloom on the prospects of the issue. As was to be expected, knowledge of British Telecom as a company amongst the City was extremely low because of its position in the public sector and therefore outside the normal interests of the market. Indeed, 61 per cent of institutional investors polled knew little or almost nothing about British Telecom's management, divisional structure or commercial activities. Consquently the measurement of familiarity of British Telecom in their professional capacity was low. Stemming from the low scores on familiarity, favourability towards British Telecom was also low with some 26 per cent of institutional investors being positively unfavourably disposed towards the company. On the twin measurements of familiarity and favourability, British Telecom was near rock bottom when compared with listed companies, the orbit of the market. Furthermore, on that crucial factor in making an investment decision, an assessment of the quality of management, British Telecom again scored badly. Not only was there almost complete ignorance of who was on the management team of British Telecom, the anonymous management was not well regarded, with an unacceptable 32 per cent of those surveyed regarding British Telecom's management as below average and 49 per cent regarding it as only average. These four crucial factors of low familiarity and favourability of BT, a negative assessment of British Telecom's management combined with its low profile were major stumbling blocks to the success of the issue. It therefore presented the enormous task of educating and changing perceptions, and of presenting British Telecom's management to around 98 per cent of institutional investors. On top of this there were the City's natural concerns about subscribing to the issue itself, based on possible Government regulations requiring maintenance of perceived unprofitable activities such as public telephones, free emergency services and directory enquiries, together with restrictions on profit growth foreseen in the pricing formula. But there were some grounds for optimism. All three key audiences, the institutional investor, specialist stockbrokers and the financial journalists were unanimous in judging British Telecom's monopoly as a key strength which combined with its commercial activities in the information and telecommunications industry and its inherent potential for growth offered opportunities of good profits and return on investment. On future prospects, providing the terms of the issue were reasonable (and it was Kleinwort Benson's essential task together with lead broker Hoare Govett to find terms acceptable to both the market and the vendors) there was a glimmer of hope, for familiarity and favourability could be increased and the management profile raised.

And this was the specific task of the second and third elements of the strategy, reinforced by a superbly orchestrated public relations campaign and underpinned by the strategy of segmenting three principal markets for the issue—i.e. the general public, the traditional UK investment institutions and the overseas investors in New York, Toronto, Tokyo, Switzerland and the rest of Europe.

Thus, the second element of the marketing campaign was to increase knowledge and awareness of the company. This was amply filled by initiating a major programme of stockbroker research work on the company. Leading the field were three reports produced by the lead brokers. The first and most comprehensive, representing something of an omnibus survey of British Telecom, was produced by Jack Summerscale and Chris Wells of De Zoete & Bevan and was launched with a fanfare of publicity at a press conference on 1 July 1984. The reports by Scrimgeour and Hoare Govett were shorter and more selective in approach but all three together represented the first detailed descriptive and analytical probe into a company that was to come to the market in the autumn. In essence the reports represented a major educational and informational building block to acquaint the market with this hitherto unknown entity and they were supported by numerous broker circulars based on seminars hosted by British Telecom. These broker reports were distributed to clients, financial journalists and opinion formers of all kinds, and above all to fund managers and investment analysts.

The early and timely provision of information to target markets was a feature of the campaign, upheld by the novel concept of issuing a Pathfinder prospectus on 26 October. As already explained, this was similar to the actual prospectus due on 20 November with certain essential information omitted (for example, the price of the shares and the dividend yield). Thus institutional investors and financial journalists had time to absorb the details contained in the 64 page prospectus and guess the final price. On 27 October, the day after the Pathfinder press conference, the *Daily Mail* correctly predicted a price of 130p a share. In essence the educational programme directed at all three markets eliminated surprise and provided time to absorb information.

The final element in the UK institutional marketing programme was the campaign to increase the profile of BT's management in the City. For this purpose British Telecom board members ran a programme of lunches from the summer of 1984 with fund managers and stockbrokers. Seminars were held to brief investment analysts and stockbrokers on the history and scope of the business. Visits to Martlesham and to the Guildhall Exhibition were arranged in order to demonstrate British Telecom's technological prowess to the City. These meetings provided an opportunity for the institutions and their advisers to get to know the company and its management and were successful in changing the City's apathy and perceptions into live interest and enthusiasm for the shares.

However, while the courtship of the investing community was crucial to the flotation, it was also Kleinwort Benson's responsibility to safeguard the interest of the Government, and to ensure that the vendor would receive the maximum proceeds from the sale, two strategies foreshadowed in the Pathfinder prospectus were employed. The first, as is usual in an offer for sale, was the underwriting operation. Coinciding with the announcement of the price of shares on Friday 16 November at 8 a.m. began the largest underwriting exercise in the market's history. In just three hours, 2.6 billion British Telecom shares were underwritten and sub-underwritten by around 2000 institutions, a process eased by the 1.5 per cent commission on stock institutions took on a firm basis and $1\frac{1}{4}$ per cent, the normal commission, on the amount they underwrote and a guarantee from the issuing house that underwriters would immediately receive 55 per cent of the shares they promised to buy instead of waiting until the lists opened. This combination of underwriting and a firm placing of the shares is novel, and was a device to avoid the double-selling to the institutions which had occurred in past offers for sale (that is, selling the shares once at the time of underwriting and again some eight days later at the time the subscription list closed). The firm-placing technique was essential given the size of the offer and the need for institutional support.

THE ROADSHOWS

The launch of the Pathfinder prospectus in London on 26 October marked the start of a major roadshow programme requiring two teams of directors drawn from British Telecom and Kleinwort Benson, together with Warburg and other key advisers to travel throughout the UK, covering 18 locations from Plymouth in the south to Aberdeen in the north as well as overseas to USA, Canada, Switzerland, W. Germany, Holland, France and Japan. The object was the essential one—to market British Telecom shares and to ensure that the overseas investment community and their advisers were as fully informed as was practicable about British Telecom and the offer prior to making their decision. Both teams would give similar speeches at home and abroad and as far as possible the answers to questions were standardised.

The objectives of the UK roadshow were almost identical with those of the BT train and it addressed itself to updating information for provincial intermediaries and fund managers about the terms of the share offer as well as the logistics for applications. The roadshows were also designed to cement the distribution network in the provinces to the retail market. They provided a further opportunity for press relations, widespread exposure in the provincial press, and an opportunity to present the management of British Telecom and at the same time obtain some feedback on the likely level of interest.

At each presentation there was a large audience of between 150 and 200 people, with interests ranging from those of the stockbroker investment analyst and institutional investment manager in London and Edinburgh to those of the provincial financial adviser. Each session was hosted and introduced by the regional co-ordinator, and began with a cut-down version of the 'Power Behind the Button' film. For 'Team A' Sir George Jefferson, chairman of British Telecom, began the presentation with an up-beat talk on the company's three principal objectives, namely improvements in quality of the network, products and management, growth opportunities offered from market penetration, increased usage and diversification of products and markets and finally profitability, a reward of growth, cost management and diversification. Continuing these themes, Iain Vallance, Managing Director of Local Communication Service division, and perhaps provider of services most familiar to the consumer, gave a short guided tour around each of BT's operating divisions noting achievements and opportunities. Deryk Vander Weyer, deputy chairman of British Telecom, began on a note of reassurance about potentially prejudicial factors such as OFTEL, the licence agreement and the pricing formula RPI−3. The regulatory system and the relationship with Government were in fact 'relatively mild and favourable to our growth'. As to the financial picture, Vander Weyer explained the reasons for the small 4 per cent decline in profitability for the year ended 31 March 1984, but stressed that substantial profits had still been earned at just under £1000 million and that forecasts for the current year would exceed £1350 million. The high quality of earnings and potential for growth, capital structure, gearing and revenue trends indicated the fundamental soundness and prospects for this investment opportunity. Each presentation was followed by a half-hour question-and-answer session and a more informal discussion period over refreshments.

In common with the PLC train exercise, the roadshows were the subject of meticulous planning and organisation to ensure that the teams and their technical support were in the right place at the right time, involving double-

Table 9.1 *Pathfinder roadshows*

1984		'Team A'	'Team B'
9–11 October		Japan	—
29	(a.m.)	Manchester	Plymouth
	(p.m.)	Liverpool	
30	(a.m.)	Zurich	Bristol
	(p.m.)	Geneva	Cardiff
31		London	Birmingham
1 November	(a.m.)	Edinburgh	Henley-on-Thames
	(p.m.)	Glasgow	
4		Departure to Chicago	Departure to Aberdeen
5		Chicago	Aberdeen
6	(a.m.)	Toronto and Montreal	Newcastle
	(p.m.)	Flight to Boston	
7		Boston	Leeds
8		New York	Norwich
9		San Francisco	Leicester
10		San Francisco	
11		Ibid/Los Angeles	
12		Los Angeles	
13		Return flight	Paris
14		—	Amsterdam
19		London	Canterbury
20		London	Bournemouth
21		—	N. Ireland
22		Germany	

Source: British Telecom

bookings for travel arrangements (see Tables 9.1 and 9.2). If they were to have an audience it was also necessary to ensure the posting of invitation cards, the distribution of information packs, and the identification badges for guests. There was the booking of venues for the conference, accommodation for speakers, the booking of caterers and transport to be thought of, and for each presentation there was the all-important stage setting for audio-visual equipment, and the display panels featuring the major terms of the offer, British Telecom, the advertising support and the prospectus.

The prestige locations, matched by a prestigious team supported by display, film and slides served to create a sense of occasion, and even after many presentations (as one participant commented) 'we still sounded reasonably fresh'. At a budget of around £$\frac{1}{4}$ million the roadshows marketed the BT issue at home and abroad.

OVERSEAS INSTITUTIONAL MARKETING

The sale of British Telecom was always regarded as primarily a UK offer with both the marketing campaign and price formulated to achieve a success in the UK. However, early in 1984 overseas interest in the marketing of British

Table 9.2 *Pathfinder roadshow sample logistics — 'Team A'*

Monday 29 October	Manchester and Liverpool
9 a.m.	Rehearsal
10.30	Presentation at Manchester, Portland Hotel
11.45	Refreshments
Noon	Press briefing and BBC and Granada TV interviews
13.00	Depart by car for Liverpool
15.30	Presentation at Liverpool, Holiday Inn
16.45	Refreshments and drive to Liverpool Airport: check in 18.00, departure 18.30 for Zurich (flying time 2 hours) then to hotel
Tuesday 30 October	Zurich
8.00 a.m.	Meeting with Swiss Bank Corporation
9.00	Meeting with Union Bank of Switzerland
10.00	Meeting with Credit Suisse
11.15	Arrive at Zunfthaus zur Meissen for presentation
11.30	Presentation starts followed by questions
12.30	Cocktails followed by lunch
14.15	Meeting with local press
14.45	Leave for airport
15.15	Flight to Geneva
16.45	Arrive at hotel for presentation
17.15	Presentation and questions
18.15	Cocktails
19.30	Flight to Heathrow (London) flying time 1½ hours

Source: British Telecom

Telecom shares was demonstrated by the appointment of Morgan Stanley & Co. in New York, McLeod Young Weir Ltd. in Toronto, Nomura Securities Ltd. in Tokyo and the Swiss Bank Corporation. The decision to sell British Telecom shares abroad was based upon two major factors. Firstly, if this issue of unprecedented size was to be a success it needed to appeal to as wide an audience as possible—an overseas sale would create much-needed additional distribution capacity. Secondly, if British Telecom was to play an important role overseas it needed to have some support among international investors as well as among the UK financial community. For these reasons offerings in Canada, Japan, Switzerland and the USA were added to the transaction, although taken together they finally accounted for only some 14.0 per cent of the shares on offer. These overseas offerings did, however, add capacity by creating interest among investors outside the UK, in turn encouraging UK institutional investors to believe that they would not get the amount of stock they had originally anticipated and indeed that there might be a shortage. This strategy of creating the perception of demand in excess of supply was one of the major factors in the ultimate success of the British Telecom issue.

The British Telecom share issue represents a genuine simultaneous cross-border offering based on the precedent set by the British Petroleum issue in 1977 which was made simultaneously in London and New York and followed much more recently by Reuters. However, any cross-border offering is subject to different practices and regulatory requirements. A major difference between the UK and US system centred on retail marketing. In the UK, before British Telecom, little effort was paid to actively selling new shares to a wide audience. On the contrary, thanks to the tax-efficiency of investments via pension funds, insurance companies and unit trusts and the attractions of building societies private shareholders had declined in number and importance.

In a typical UK new issue offer the prospectus and the application form must be advertised in at least two national newspapers. This does provide a very simple opportunity for the retail investor to subscribe but it is at best a passive system. Normally, no selling commissions are paid and there is consequently little or no direct selling effort by stockbrokers to the retail sector, except perhaps to a small number of private clients. It was estimated in 1984 that only some 3 per cent of the population were acquainted with a stockbroker. There is in effect a complete absence of any kind of active retail share distribution system. Another feature of new issue sales is that institutional marketing is confined to London and Edinburgh with little attempt to spread interest nationally. In this situation it is very difficult for the issuing house to assess retail demand and, in most UK offerings, a high level of retail demand is unnecessary.

The US situation is very different. The major retail brokerage houses have offices throughout the country to ensure national retail distribution of shares. Selling commissions of around 3 per cent are common to reward the sales efforts of teams of equity salesmen. The lead manager has available through the underwriters' retail brokerage network a developed mechanism for assessing the level of retail demand prior to the pricing of the offer. The different market approach between the UK and USA is usually summed up in the old adage: 'In the UK investors buy securities, in the US they are sold securities'.

The second major difference is between underwriting and offering practices. In the UK, issues are underwritten by a merchant bank or a group of them immediately before the announcement of the price. The issue is immediately sub-underwritten with a group of institutions on the first day of the public offering. The public offering is then open for a period of approximately six days for the public to subscribe. Trading will start perhaps four days later making a total of ten days during which the underwriters are at risk.

The US practice is very different. On Wall Street a draft prospectus known as a 'red herring' (model for the British Telecom Pathfinder prospectus) is circulated among prospective investors simultaneously with the Securities and Exchange Commission's (SEC) review of the documentation. During this period the lead-manager and underwriters arrange presentations (or roadshows) across the country and 'build a book' indicating the level of interest from institutional clients and from retail equity sales teams. In the light of the response of the investment community to these presentations and the prospectus, the leader of the underwriting syndicate in discussion with the issuing house may then vary the size of the issue, or its price, or both in an attempt to satisfy the level of indicated demand. The issue is then priced, generally late in the afternoon after the market has closed and underwriting commitments are made. The next day, the shares are sold by the underwriters to the institutions and through the brokerage houses to the public. Trading starts immediately so that those who have underwritten the deal are not exposed to fluctuations in the share price. In essence therefore UK underwriters hold securities and are at risk for many days; in America underwriters are purchasers and distributors of shares taking a risk for only a matter of hours.

There is also the different regulatory environment. This extends beyond the different level and type of disclosure that is required under Stock Exchange and SEC rules. In the UK, the Stock Exchange sees all offering documents, draft by draft, so that no document is published without approval. However, recent EEC Directives have been incorporated in The Stock Exchange Yellow Book (otherwise known as the 'Yellow Peril'!) which allows very little flexibility, and the British Telecom prospectus was designed to comply with these. The SEC by contrast is not part of the New York Stock Exchange, rather it is a Government agency enforcing a relatively tight series of rules established by legislation with only limited discretionary powers. This difference is reflected in the respective roles of financial and legal advisers. In the UK, the drafting of the prospectus is the sole responsibility of the issuing house assisted by lawyers. All matters that involve the Stock Exchange are handled by the brokers to the offer. Conversely in the USA the legal advisers have a more important role in the drafting of the prospectus and generally handle matters with the SEC.

There are also differences in the accounting practices of the two markets requiring different types and levels of disclosure. Also significant are the very different liabilities that arise particularly for company directors in connection with the preparation of the prospectus. Given these major differences between financial centres and capital raising techniques it is obvious that simultaneous cross-border equity offerings require substantial co-ordination to meet not only regulatory requirements but also market practice, including the provision, registration and filing of separate documentation. Thus registration was undertaken with the appropriate authorities of the BT prospectus in America, Canada and Japan. On 26 October the 'red herring' prospectus was filed in America and Canada. The following day a preliminary prospectus was filed with the ministry of finance in Japan. These 'red herring' documents, though similar to the UK Pathfinder prospectus, differed in layout in order to meet local custom and local regulatory requirements. Also, to cover the gap in time in underwriting between the UK and overseas, particularly in the USA, the Bank of England underwrote the overseas portion of shares, a solution that is clearly not available to private sector companies.

Table 9.3 *Timetable of UK and USA British Telecom share offer*

	UK	USA
26 October	Publication of UK Pathfinder	Publication of US 'red herring'
26 October to 15 November	Marketing	'Building the book'
16 November	IMPACT DAY	
	Fixing of price and signing of underwriting agreement with UK syndicate	Republication of 'red herring' with price included
20 November	Publication of final UK prospectus with price	
20–28 November	Offer open to general public in UK	Marketing
28 November	UK offer to general public closes	Signing of underwriting agreement with US syndicate
3 December	Dealings commence	Dealings commence

The timetable of the simultaneous offering of British Telecom shares in the UK and USA can be seen above (Table 9.3).

The period of the issue itself and its final outcome is well documented above. A Pathfinder prospectus, containing full details of the company together with a profits forecast to March 1985, was issued on 26 October 1984. Three weeks later on Friday 16 November the issue was priced and underwritten by a syndicate of 15 UK banks. The final prospectus was published on 20 November. The issue was then held open for eight days during which time it was widely advertised in the press, on television and radio. The overall response both in the UK and overseas exceeded the earlier expectations of the organisers.

Chapter 10

Afterthoughts

On Tuesday 20 November 1984 the British Telecom offer for sale for 3012 million shares worth some £3916 million was formally launched with the publication of the full prospectus of some ten pages in four national newspapers, the *Financial Times, Guardian, The Times* and the *Daily Telegraph*, accompanied by advertisements for the 'mini' prospectus and application form in the *Daily Mail, Daily Mirror, Daily Star* and the *Standard*. These advertisements were reinforced by additional publicity on television, radio and posters. Simultaneously, for the 1.3 million who had contacted the British Telecom Share Information Office (Bristol) for further details, the 'mini' prospectus dropped through over a million letter boxes and for those visiting their post office or bank branch the 'mini' and full prospectuses were waiting on counters. On Wednesday intense publicity continued with six newspapers carrying prospectus advertisements. The countdown to the closing date had begun amid the glare of intense and widespread publicity accompanied by a crescendo of press advice that this offer was simply too good to miss.

The application lists opened at 10 a.m. on Wednesday 28 November and closed a minute later for the start of the biggest share count in history, recorded the next day by an official photocall at each of the six receiving banks, Barclays, Midland, NatWest, Lloyds, Royal Bank of Scotland and the Bank of Scotland. It showed to an expectant and hopeful public the counting clerks at work sorting applications and processing cheques. By Friday lunch-time the registrar's department of Lloyds Bank had the final figures for the total number of applications broken down by type—British Telecom employee and pensioner applications, numbers opting for the telephone vouchers, and those choosing the bonus issue. The final results were communicated to the Prime Minister (then in Paris) and to Kleinwort Benson. Early on Saturday 1 December the allotment committee met to agree the basis of allotment, mindful of the Prime Minister's desire that small investors should receive priority.

The issue had been substantially oversubscribed, with around £16575 million chasing £3916 million worth of shares on offer. Demand in the three

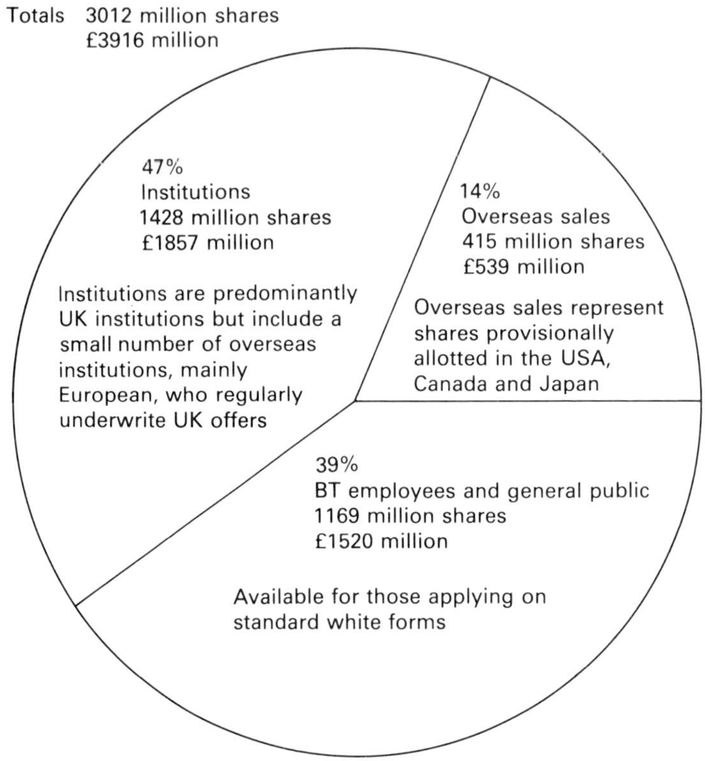

Figure 10.1 *British Telecom share offer.*

markets exceeded supply. The public had applied for more than five times the number of shares set aside for it. The UK institutions found themselves underweight in a key stock and the overseas sector had been scaled down. The pie chart (Figure 10.1) shows the final allocation of shares between the three markets. The UK financial institutions accounted for the bulk of the shares at around 47 per cent of the total number of shares. The general public, including British Telecom employees, took 39 per cent of the offer and the remaining 14 per cent went to America, Canada and Japan. Given the substantial demand from the public all applications were scaled down in favour of the small investor as can be seen in Table 10.1.

The one million people who had applied for between 200 to 400 shares received the full amount applied for, whilst the 1066000 applications for over 400 shares were heavily scaled down with those applying for between 1600 to 100000 shares receiving a very small number of shares and those who applied for over 100000 shares receiving nothing. A more precise picture of British Telecom's retail shareholder profile (excluding BT employees) before dealings began was given by Mr. John Butcher, Under-Secretary of State for Trade and Industry, in reply to a Parliamentary question (Table 10.2).

Table 10.1 *Allocation of shares in the retail public offer*

Application for shares (number)	Allotted shares
200–400	Granted in full
800	500
1200	600
1600–100 000	800
Over 100 000	Nil

Source: Kleinwort Benson

The largest share issue in the world was a substantial achievement and a great success: the *Sun* declared, 'Take off. The sale of British Telecom is a brilliant success, with ownership of the company spread among 2 million investors'. The objective of a full subscription to the offer had been more than met. The Government's desire for wider share-ownership had been delivered by the retail marketing drive. The privatisation of British Telecom had also given a considerable fillip to the Government's drive on the employee shareholder front with over 220 000 BT employees seizing the opportunity of the free offer, a feature noted by John Moore, Financial Secretary at the Treasury in a speech to the Institute of Directors in December 1984: 'One of the many remarkable features of the BT sale is the number of BT employees who have subscribed for shares'.

At 3.00 p.m. on 3 December dealings in British Telecom shares began in London and New York. The shares made a sensational debut in the London stock market opening at a premium of 45p over the first instalment of 50p, well above expectations, and investors at the close of trading enjoyed a paper profit of nearly 90 per cent on their original investment. The price of BT shares since their stock market debut has risen to a peak of 206p on a 90p paid-up share. The splendid performance of the shares in London (Figure 10.2) has provided a welcome introduction to equity investment for millions of first-time buyers.

Table 10.2 *BT Retail shareholder profile on 29 November 1984*

Number of shares	Number of shareholders
0–200	525 698
400	512 063
500	335 945
600	117 501
800	625 971
1200	3 461
1600	12 706
over 1600	8 302 (of which 2127 are institutions)
Total	2 141 647

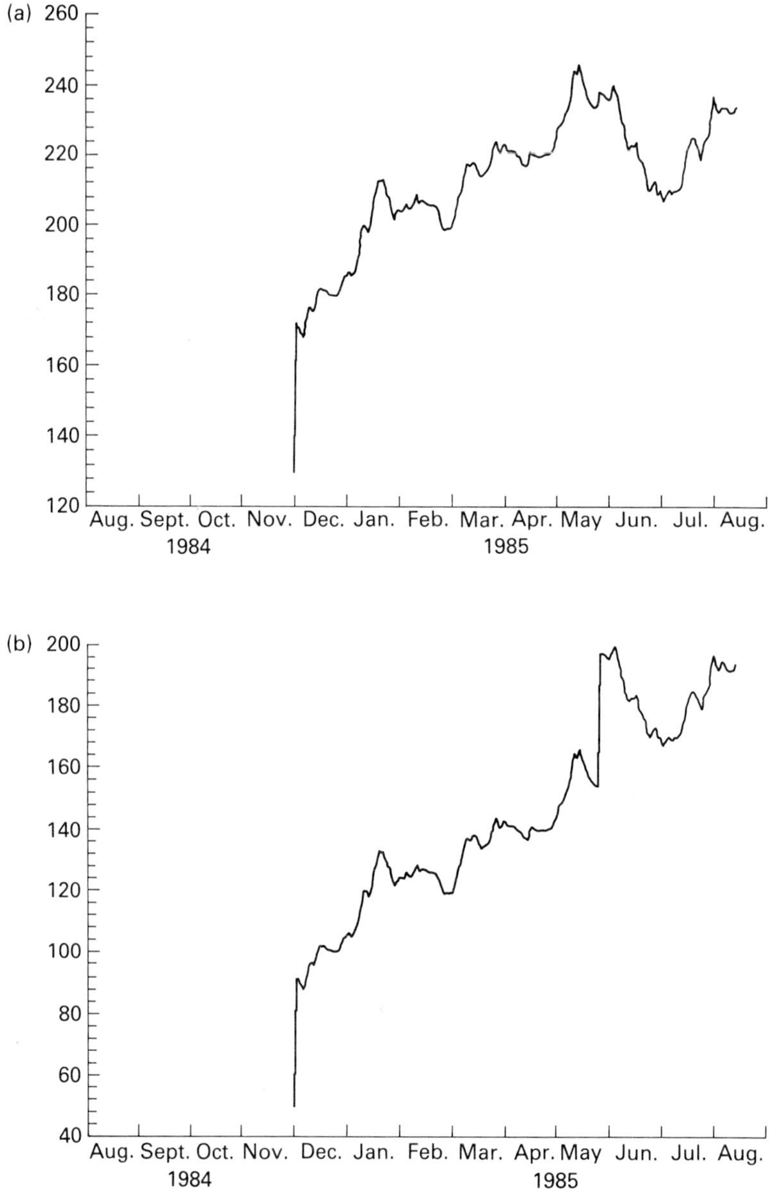

Figure 10.2 *BT shares: (a) fully paid; (b) partly paid.*

By the end of May 1985 around 450 000 people had sold their shares leaving around 1.69 million on the BT share register thereby confirming predictions based on MORI research that the majority of investors were long-term shareholders. However, such sales were sufficient for the BT share register to exhibit the normal pattern of a 70:30 split in favour of the institutions. Thus in

June 1985 it is estimated that private individuals accounted for a mere 29 per cent of British Telecom's shares, compared with 39 per cent at the outset, the remainder being in the hands of the customary investors in equities, the financial institutions. Research by Taylor Nelson Financial, shown on Channel 4's *Business Programme* on 26 June 1985 shows a markedly uneven geographical distribution in BT share-ownership. Not surprising perhaps is the high take-up in the south east of England, which accounts for 42 per cent of the total, and the 30 per cent in wealthy East Anglia. It is the low take-up in the north of England and Scotland that is surprising, with only 12 per cent of the shares, and the relatively low take-up in the West Country with only 14 per cent. Characteristically, men outnumbered women and it is clear that BT shareholders show a significant increase in share-ownership among the C2s and those under 54 years of age.

The success of the marketing campaign lies in achieving a full subscription to the largest flotation in history and attracting substantial numbers of first-time buyers into the share market, thereby doubling the private shareholder base of the UK. Whether this is the beginning of a trend is difficult to predict. Certainly there is some anecdotal evidence to support an increase in awareness and interest in equities among the general public. Abbey Life, the UK's second largest unit-linked life company floated in June 1985 was 19 times oversubscribed with some 375000 applicants and £4 billion chasing shares worth £241 million. Similarly the final tranche of British Aerospace was oversubscribed more than five times and that of the Government's remaining 49 per cent stake in Britoil was four times oversubscribed overall, or ten times oversubscribed in the 15 per cent of the issue reserved for the general public. Brokers report an upsurge in general interest in shares and telephone calls to newspaper offices indicate interest from lower down the socio-economic scale. Encouraged by the success of the marketing campaign, lead stockbrokers to the British Telecom issue, Hoare Govett, have introduced a low-cost, minimum-frills dealing service called 'Dealercall' which offers speed and simplicity for deals worth some £750 and a number of brokers particularly outside London are making themselves more available to members of the public and have reduced their expectations as to the size of individual investments. But how many of the people who invested in BT or are employee shareholders have subsequently dealt on the stock market? MORI research indicates above average interest by the general public in forthcoming privatisations such as TSB and British Gas but this, rather than reflecting a general interest in shares *per se*, may reflect a narrow interest in one kind of stock—that sold by the Government. The very success of the British Telecom flotation may have merely created a privileged sector of equities—state stocks which through favourable pricing and the vendors' credentials carry a special 'clean bill of health'.

The Government's expectation of wider share-ownership, based on the American example where 25 per cent of adults own shares, is questionable. The UK pattern of saving and investment profile reflects differences in tax structure and national wealth. The UK tax structure favours channelling savings through pension funds and, until the Budget of 1984, through life assurance schemes. In America the pensions sector is small and companies have regularly used employee shares as compensation to employees. In America there are many wealthy individuals for whom it is appropriate to hold substantial sums in risk-bearing securities and the USA has a larger pool of private investors who have the skill and experience to make their own portfolio decisions. The population of Britain is relatively poorer, less sophisticated and lacks the cheap retail share distribution network of America. Only time will tell whether individual share-ownership in the UK will become analogous to home ownership. Certainly the achievement of British Telecom cannot be ignored but nor can the scale of publicity be imitated by private sector companies coming to market. Few companies could ever afford the marketing costs of this offer with its unparalleled advertising, retail distribution network, and generous commission payments to financial intermediaries. Few companies coming to the market will have a ready-made mailing list the size of British Telecom's customer list. Nor will they have the advantage of regular correspondence with their customers. And throughout the planning and execution of the offer campaign there was the bottom line assumption that the offer would be attractive. Although the planners of the flotation were conscious of the daunting problems which faced them at the outset, few companies in the private sector could hope to acquire such outstanding advantages that British Telecom eventually acquired.

> Without further stimulus, it would be unwise to assume that the BT flotation marks a new dawn in the way the nation conducts its affairs. This was an offer not to be refused, priced at an enticingly low level, baited with vouchers to be used against telephone bills and virtually underwritten with a promise of immediate profits . . . Not for nothing has the BT issue been dubbed 'Sale of the Century'.
>
> The *Yorkshire Post*, 3 December 1984

However, several questions remain unanswered concerning the marketing campaign. Should people be asked to subscribe to shares through direct mail, for it is a dangerous weapon in the hands of the unscrupulous and until it was outlawed in 1938 it had an unsavoury reputation. Whilst it is notoriously difficult to price any new issue the BT flotation was complicated by its sheer size. To sell 51 per cent of BT in one tranche, combined with the political need for success, tended to lead to a low-pitched price. Should it not have been higher? The very success of the public relations campaign is awesome in its implications for it virtually captured the press, and is almost alarming in its power.

In a series of written Parliamentary answers, Mr. Butcher, Industry Under-Secretary, suggested that the offer advertising and marketing costs reached some £17.5 million. This figure includes £7.6 million for the 'Alert', 'Maintenance' and 'Action' phases, the prospectus advertising of around £3 million and other marketing costs, such as printing literature and the prospectus, of around £6.9 million. However, the Comptroller and Auditor General in July 1985 put the UK marketing cost at £14 million and total expenses at £263 million. This is a higher proportion of proceeds than in all but one of the six previous sales of Government shareholdings where costs on a comparable basis averaged 3.3 per cent of the gross proceeds.

With hindsight the marketing strategy is vulnerable to criticism. The scale of the oversubscription by the retail sector suggests an overkill in advertising and promotional literature. At the same time it has to be remembered that, perceived in advance, such an outcome would have seemed fantastic. Indeed, the tone of the whole campaign had been educational, designed to inform rather than make a 'hard sell', leaving the decision to the public. But, despite this constraint, there was a touch of 'Madison Avenue' about the campaign and the sheer sophistication of glossy literature directed at the public achieved a level of showmanship more suitable to selling consumer products. British Telecom was sold only on the future, with a one-year profit forecast to March 1985. It had no track record of commercial judgement and offered only a promise rather than proven ability to control operating costs.

Furthermore, it is questionable whether the simultaneous cross-border offering, insisted upon by the UK institutions, was a good strategy. Whilst there is a legitimate argument for BT to have international shareholders the portion sold in the USA and Canada returned almost immediately to London, although the number of shares held in Japan has increased from around 180 million at the time of the flotation to around 200 million six months later, a reflection of Japan's overseas portfolio investment as much as the attractions of BT shares. Whilst the success of British Telecom in the domestic market seems assured, exponential growth opportunities lie in global marketing and information technology areas. It is in these areas that British Telecom is vulnerable. British Telecom has not yet proved that it has the management ability to achieve significant foreign market entry or to survive, let alone compete, in a global market-place. Furthermore, in the information technology sector the rapid technological convergence of computers and telecommunications brings BT into direct competition with multinationals like IBM of America and Siemens of West Germany.

There are also other considerations such as the somewhat lax regulatory structure created by the Telecommunications Act 1984 in which marketing requirements appear to have outweighed the Government's declared competition policy. The questions abound. Was it for example sensible to take the

Office of Fair Trading rather than the Civil Aviation Authority or some other regulatory body as a model? By comparison with American regulatory authorities it was clear from the outset that OFTEL would offer no more than the minimum acceptable degree of control, since it was lacking in both sanctions and resources. The Director General lacks the power to license new competitors to British Telecom and has a very modest staff with which to follow up complaints of anti-competitive practices. In his first report, the Director General notes the difficulties of investigating alleged abuses of market power because complainants wish to remain anonymous. This desire for anonymity is suggestive of timid David ducking from a more formidable Goliath. Equally OFTEL's duty to supervise BT's compliance with the provisions of its licence is fraught with difficulties. The terms of BT's licence require it to provide figures for the profits of its network and profits from its equipment business. This segmental information is the key to assessing whether there is any cross-subsidisation of the two services. The terms of the licence lay down that this segmental information should be provided as soon as is practicable and not later than 1 April 1987. BT's first set of report and accounts as a public quoted company are devoid of this crucial information. The absence of the figures makes OFTEL investigations of allegations of cross-subsidisation against BT virtually impossible, thereby placing OFTEL's authority in doubt. Equally important, BT's reticence or inability to provide such figures conflicts with Sir George Jefferson's claim in his chairman's statement that he wants BT to be a 'top class information technology business operating around the world'. Such information is as crucial to BT when planning its corporate strategy as to OFTEL to restrict monopolistic practices. The two roles of regulation and consumer representation in one body, OFTEL, appear to be difficult to reconcile.

Moreover the macro-economic effects of the privatisation of British Telecom require some consideration. The Government's declared objectives of competition and efficiency criteria required the breaking up of British Telecom. The American example of AT & T, which was forced to divest itself of its local companies has shown very considerable savings in manpower and an increase in efficiency. There is no such stimulus available to BT, and for the foreseeable future, no network can be expected to rival British Telecom's capacity and coverage. For most consumers British Telecom is the monopoly provider of a basic service and for the majority of them it will continue to be so, even when rivals such as Mercury and cable television appear as alternative carriers. The privatisation of BT has merely exchanged a public monopoly for a private monopoly with weak competition at the periphery.

Does privatisation result in greater efficiency? It is too soon to give a verdict, but the removal of restraints, which is the market rationale for privatisation, has yet to prove its value. The evidence so far suggests that privatised companies have simply sustained improvements that were already taking place while they were still in the public sector. Yet over the two years during which British Telecom had to prepare itself for private ownership there was a marked shift in BT's responsiveness to its customer base, changed attitudes and improved performance. Under the prospect of privatisation BT's waiting list which exceeded a quarter of a million in 1979 was down to 2000 in 1984. Likewise three-quarters of residential orders were fulfilled within 12 working days and 85 per cent of faults were cleared by the next working day. Efficiency lies in good management rather than ownership. The real question is 'Why can't we have an efficient well-managed public enterprise?'

The question whether the Government is simply selling off the family silver to finance income tax cuts in time for the next general election remains. Privatising 51 per cent of public corporations removes them from the public sector borrowing requirement and boosts Government revenue in the short run. On the evidence so far, over-attractive pricing of shares means that Government receipts from such offers for sale have fallen well below any reasonable measure of the discounted stream of future income from former state industries. Moreover, the argument that privatised companies would in future tap the capital markets for additional funds instead of the Treasury is not so foolproof either. Newly privatised companies may find it difficult to raise fresh money within three years of public flotation and future planned privatisations worth an estimated £8.5 billion over the financial years 1986–87 and 1987–88 may 'crowd out' company financing through equity issues.

Nevertheless, Government ministers have described privatisation as 'among the most radical reforms of the UK's economic and industrial structure since 1945' and since 1979 14 major enterprises have been privatised, around 400000 employees have been transferred from the public to the private sector, around a third of a million employees have acquired shares in the companies for which they work and state sell-offs have raised over £6 billion for the Exchequer. The successful privatisation of British Telecom marks a qualitative breakthrough in the formation of public opinion and perception. By attracting over 2 million investors the flotation appears to have removed remaining inhibitions about privatising publicly owned enterprises. In July 1985, the Government announced an accelerated programme to privatise eight enterprises, including natural monopolies like gas, water and electricity which will raise as much money as in the previous seven years and reduce the share of state industry in the gross national product from 10.5 to 6.5 per cent.

Index